CHLORELLA

THE EMERALD FOOD

DHYANA BEWICKE

BEVERLY A. POTTER, PHD.

et al

RONIN PUBLISHING

BERKELEY, CA

Chlorella: The Emerald Food
ISBN: 0-914171-02-X

Publishing by Ronin Publishing Inc. Berkeley, CA

First Printing: November 1984 Current printing 10 9 8 7 6 5 4 3 2 1

Library of Congress Cataloging in Publication Data

Bewicke, Dhyana.
Chlorella : the emerald food.

Includes bibliographies.
Contents: The chlorella story / Dhyana Bewicke -- The amazing alchemist / Beverly A. Potter.
1. Chlorella as food. 2. Chlorella. 3. Algae culture. I. Potter, Beverly, A. II. Title.
TX402.B49 1984 641.3 84-22334
ISBN 0-914171-02-X (pbk.)

Acknowledgements

Project Director: Beverly A. Potter
Developmental Editor: Sebastian Orfali
Manuscript Editor: Judith Abrahms
Cover Design: Brian Groppe
Layout & Production: Brian Groppe, Beverly Potter
Data Entry Coordinator: Iris Miller
Typesetting: Generic Typography
Printing: Delta Lithograph
Technical Advisor: Sandy Szabat
Expert Advice: Dr. William Oswald, University of California at Berkeley; Dr. John West, University of California at Berkeley; Dr. Mel Avener, NASA; Carole Korb, Sun Chlorella California, Inc., Redondo Beach, California; Cal Bewicke, Laurel Canyon Chlorella
Research: Linda Finegold, Anne Moose, David Nonomura

Table Of Contents

Chlorella: The Emerald Food

Book I

The Chlorella Story

DHYANA BEWICKE

Book II

The Amazing Alchemist

BEVERLY A. POTTER, PHD.

BOOK I

The Chlorella Story

DHYANA BEWICKE

Preface

Dhyana Bewicke has written an important book in the field of health and nutrition. Her subject, the Chlorella microalgae, has become the focus of interest and excitement in this country through growing awareness of sophisticated technologies of food production, which have been developed in Japan over the past twenty-five years. In the past two decades Chlorella has become popular with millions of Japanese people; more recently, its benefits have come to the attention of Americans who are interested in nutrition and health.

One of the greatest factors that make Chlorella an important food is its extremely high level of chlorophyll, a substance necessary to good health and detoxification in people and animals alike. I would even say that chlorophyll is the most important part of microalgae, providing the most benefits. *Chlorella: The Emerald Food* explains the health-giving qualities of Chlorella, and provides much important information about its nutritional composition. The analyses of Chlorella are factual and accurate, and the conclusions based on these analyses provide a framework around which readers can build nutritional programs for themselves using this potent force for health.

Dhyana Bewicke also provides much easy-to-use information about the use of Chlorella in weight control and weight loss, as well as the importance of using Chlorella in dietary programs for children or older people.

Chlorella is a food of proven benefit, a tremendously promising food for our time. It has been tested and retested as a food supplement and as a restorer of health. It is a truly effective means of insuring health in today's polluted and stressful world.

Jeanne Rose
author of
Jeanne Rose's Herbal (Grossett and Dunlap)
The Herbal (Bantam Books)

A New Frontier in Food Production

In the past few years, many remarkable new foods have become available to health-conscious American consumers. Soybean products such as tofu and tempeh, nutritional yeasts, and many traditional healing herbs are widely used by those who seek a more natural and healthy way of life. The most remarkable New Age food, which contains the highest concentrations of whole food nutrition known to exist on Earth, are the single celled microalgae: Chlorella, Spirulina and Dunianiella.

The History of Chlorella

Microalgae were among the earliest and most primitive life forms to appear on our planet. They have resided at the very base of the food chain and have been an indispensable part of the ecosystem of Earth for about two billion years. Microalgae have been used as a source of vitamins and protein since ancient times. Early civilizations, including that of the Aztecs, used microalgae as an important part of their diet. Seaweeds (which are another form of algae) have been used in the Orient for thousands of years, and are now well known in the West for their high quality of nutrition.

The first scientifically pure cultures of algae were *Chlorella vulgaris*, grown in 1890 by the Dutch microbiologist, M.W. Beijerinck. By 1919, Otto Warburg had published articles on his use of dense laboratory cultures of Chlorella in the study of plant physiology. After years of intensive research on Chlorella and other microalgae, it became clear that microalgae, grown under proper conditions, can produce nutritional benefits more efficiently than those provided by the higher plants. Early microbiologists speculated that since algae have such high nutritive value (they contain as much as 50-60% protein), large-scale production could lead to a revolution in agriculture.

During the 1940s, two researchers, Jorgensen and Convit, fed a soup made from concentrated Chlorella to 80 patients at a leper treatment colony in Venezuela. The improvement in those patients' physical condition was the first documented evidence of the potential of microalgae as a health supplement.

It was not until the early 1950s, however, that research into the use of microalgae as a source of food and medicine for human beings began to gain momentum. This research was spearheaded by the Japanese, who began with a strain of Chlorella. The use of Chlorella as a premium-quality natural food supplement quickly caught on in Japan, where it is used daily by millions of people. Now, thirty years after serious research on this source of nutrition began, Chlorella is just becoming available to the American public.

Calloway, a renowned nutritionist, pointed out that microalgae are technologically attractive because they offer the promise of increased food production without dependence on traditional agricultural methods. Numerous studies conducted in the 1950s and 1960s in the U.S., the U.S.S.R., Japan, Germany, and Israel on the mass production of microalgae led to the conclusion, however, that microalgae were not cost-competitive with protein sources such as soybeans. This situation may be changing now. Dual-purpose cultivation, such as the work done by Dr. Oswald of the University of California (Berkeley) in which algae are used to simultaneously treat animal waste and produce animal feed and other new technologies being developed throughout the world promise to make the cost of production competitive with that of conventional food sources.

Desired Properties of Algae As a Source of Single Cell Protein

High growth rate
High protein content
Resistance to climatic variations
High nutritive value and digestibility; non-toxic
Good acceptability and palatability
Simple harvesting and processing methods
Economical production

From: E.W. Becker & L.V. Venkataraman, Production of Algae in Pilot Plant Scale: Experiences of the Indo-German Project, in Shelef and Soeder (eds), *Algae Biomass*, 1980, p.37

For about thirty years a lucrative industry producing and marketing Chlorella products has flourished in Japan and Taiwan. For example, in 1980 large amounts of Chlorella meal were sold in bulk for between $5 and $15 per kilogram, for use in pills, extracts, and other health food items. Japanese consumption of Chlorella products is over $100 million per year. Chlorella is popularly used in Japan as an ingredient of *wasabi*, the hot green horseradish eaten with *sushi*. Perhaps its most popular use is as an ingredient in fortified noodles.

During the 1960s there was a flurry of research on the use of Chlorella in outer space for producing oxygen and food. A kind of "algae space race" developed between the U.S. and the U.S.S.R. Dr. Oswald demonstrated that algae could support the entire metabolism of an adult man. His results were soon duplicated by Kondratyev and others in the Soviet Union. The results pointed to algae as an ideal food for outer space travel.

A new surge of Chlorella cultivation has been spurred by the growth of the profitable health food industry in the United States and by recent advances in enhancing its digestibility. The recent popularity of Spirulina has brought fresh attention to the unique benefits of Chlorella as well.

From its beginnings, the science of algo-culture, the process of growing and harvesting microalgae, has pursued a remarkable visionary goal: a totally new method of food production which applies advanced concepts of science to some of the most primitive organisms on our planet. In the near future, this new food source could help solve many of the problems of a hungry world. Today we have the opportunity to experience the first fruits of this vision: the tremendous benefits of the great nutritional value contained in Chlorella microalgae.

What is Chlorella?

Chlorella is a green micro-alga. Unlike more highly evolved life forms, it is unicellular which means each cell is a self sufficient organism with all the plant's life functions taking place inside each cell. This structure results in unusually high concentrations of important nutritional ingredients. *Algal Culture*, a source book of microalgae edited by John S. Burlow and published by the Carnegie Institute states:

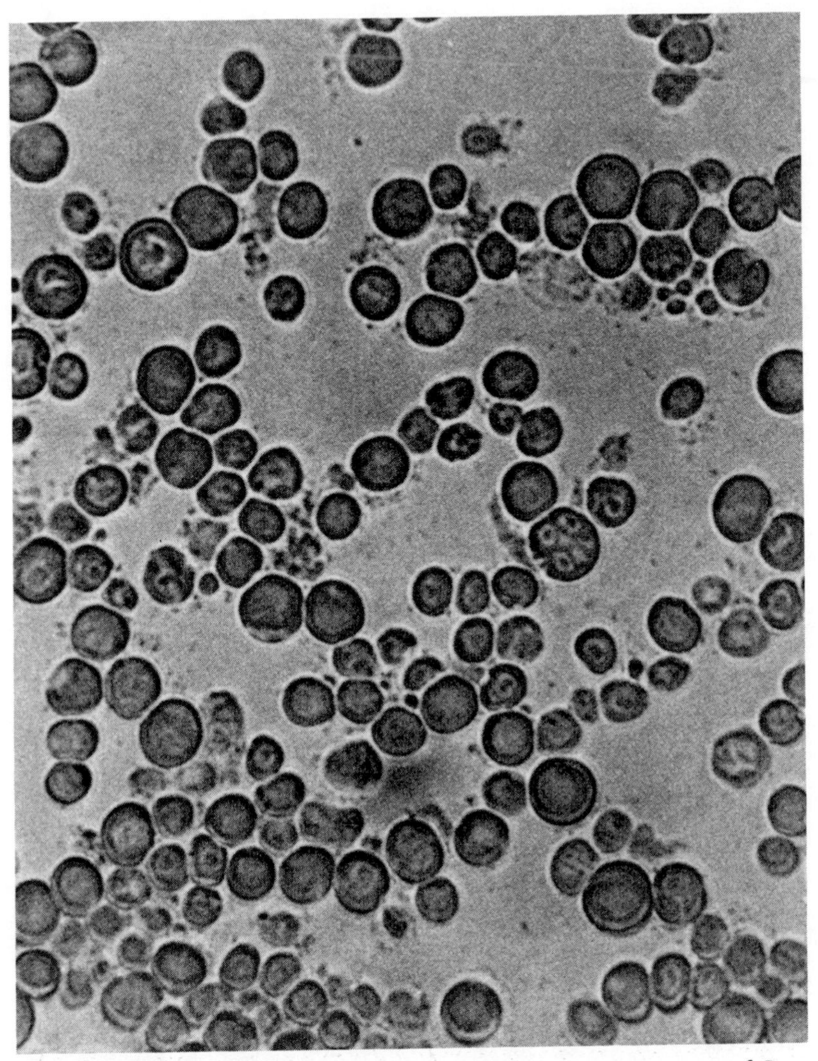

Microphotograph of purely cultured Chlorella; Courtesy of Sun Chlorella Company

In order to understand the interest in algae as a possible source of food, we must recall the general features of the structure of higher plants. Except for the leafy vegetables that are eaten because of their flavor and vitamin content, man's vegetable foods are derived from a portion of the plant, such as its fruit, seeds, or roots. These parts contain the largest concentrations of protein, which is essential for the reproduction of the plant, and of fats and carbohydrates. which are forms of food stored for the use of the next generation. The sum total of these nutritive parts of the plant, however, is usually half or less of the total dry weight. Most of the plant structure serves mechanical purposes: roots to anchor it and to draw food and water from the soil, leaves to expose large areas of cells to sunlight, and stems to support the leaves and fruits in the light and air. The primitive character of their cellular organization gives microscopic algae a number of advantages over higher plants as a source of food. In the first place, the entire plant is nutritious, for little of it is devoted to indigestible structures.

Each minute Chlorella cell measures only two to eight thousandths of a millimeter (micron) in diameter — about the size of a human red blood cell. The difference is that Chlorella is ball-shaped, whereas red blood cells are disk-shaped and about two microns thick. In fact, Chlorella cells are so small that one quart of a bright green, moderately thin suspension of Chlorella contains over 20 billion cells. While it is growing vigorously on a sunny summer day, this number of cells may easily double.

Chlorella grows in fresh water the world over. Along with other photosynthetic microalgae which produce chlorphyll and convert carbon dioxide to oxygen through the process of photosynethsis, Chlorella has played a vital role in creating the biosphere, the entire network of life of planet Earth. When microalgae first evolved, the atmosphere of the earth contained only 0.1% oxygen, photosynethic microalgae, reproducing for millions of years before the emergence of higher life forms, are largely responsible for the 20%-oxygen atmosphere in which we now live and breathe.

Today these ancient organisms, so vital to the creation of our life-nurturing biosphere, are playing a new and important role for mankind. As we look deeper into the qualities of the minute green Chlorella cell, we get a glimpse of the many unfathomed mysteries that are woven into Nature's chain of life.

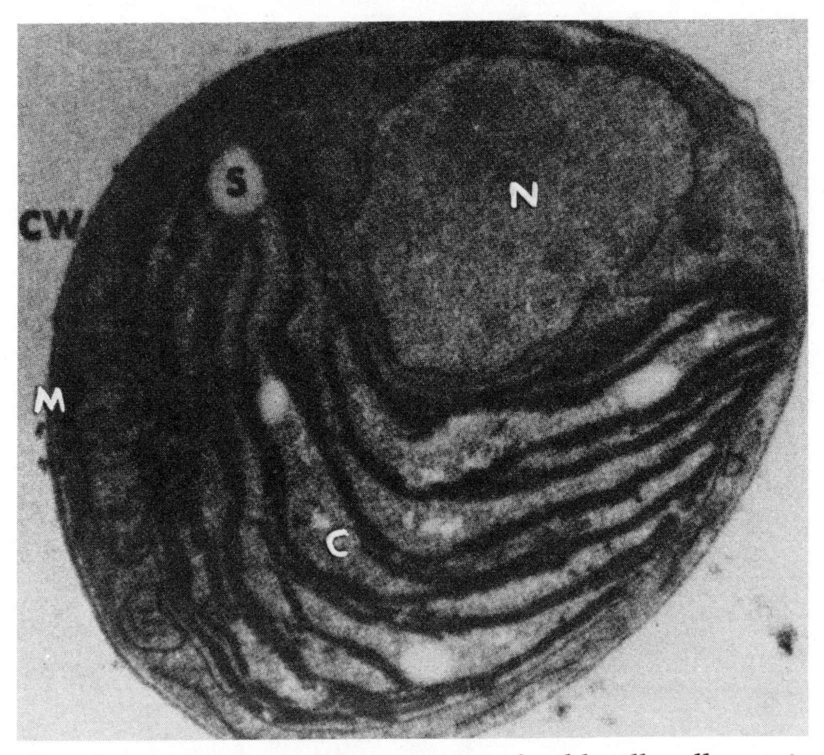

This photograph depicts the structure of a chlorella cell magnified 10,000 times. Its actual diameter is six microns. The surface of Chlorella is covered by the cell wall (CW), composed chiefly of cellulose. Within the cell can be seen the nucleus (N), starch grains (S), and belt-shaped chloroplasts (C) in which photosynthesis takes place. The mitochondrion (M) the part of the cell where metabolic energy is produced. Photo Courtesy of Sun Chlorella Company.

Chlorella as Food

High Protein Content

One of the most remarkable nutritional qualities of Chlorella is its high protein content. Chlorella is over 60% protein, including all the amino acids essential to human nutrition. This is almost three times as high as the protein content of beef, which is one of the most concentrated protein sources available. For its protein alone, Chlorella is useful as a protein powder, a food supplement, or a survival food.

In general, dried algae cells yield approximately 50% protein which is more than can be found in the edible parts of the higher plants. Algae protein, moreover, has a low molecular weight, which means that it may readily be digested, provided that the algal cell walls have been disrupted. A research group led by Dr. Dam concluded that humans are able to consume algae as their principal protein source for 20 days with no ill effects. In this study, algae was used to supply 90-95% of the protein needs of the human subjects.

Analysis of Ingredients in SUN-CHLORELLA "A" Per 100 grams

General Analysis

Moisture	3.6%
Crude protein	60.5%
Crude fat	11.0
Carbohydrate	20.1%
Crude fiber	0.2%
Crude ash	4.6%
Calorie	421 cal

Amino Acids

Lysine	3.46 w/w%
Histidine	1.29 w/w%
Arginine	3.64 w/w%
Aspartic acid	5.20 w/w%
Threonine	2.70 w/w%
Serine	2.78 w/w%
Glutanic acid	6.29 w/w%
Proline	2.93 w/w%
Glycine	3.40 w/w%
Alanine	4.80 w/w%
Cystine	0.38 w/w%
Valine	3.64 w/w%
Methionine	1.45 w/w%
Isoleucine	2.63 w/w%
Leucine	5.26 w/w%
Tryosine	2.09 w/w%
Phenylalanine	3.08 w/w%
Ornithine	0.06 w/w%
Tryptophan	0.59 w/w%

Vitamins and Minerals

Vitamin A activity	55,500 IU/100g
B-carotene	180.8 mg
Chlorophyll a	1.469 mg
Chlorophyll b	613 mg
Thiamine	1.5 mg
Riboflavin	4.8 mg
Vitamin B6	1.7 mg
Vitamin B12	125.9 mcg
Vitamin C	15.6 mg
Vitamin E	less than 1 IU
Niacin	23.8 mg
Pantothenic acid	1.3 mg
Folic acid	26.9 mcg
Biotin	191.6 mcg
Para-amino-benzoic acid	0.6 mg
Inositol	165 mg
Calcium	205 mg
Phosphorus	959 mg
Iodine	0.6 mg
Magnesium	315 mg
Iron	167 mg
Zinc	71 mg
Copper	0.08 mg

Fatty Acids

Unsaturated	81.8%
Saturated	18.2%

SUN-CHLORELLA "A" is a product of Sun Chlorella Company.

It is almost certain that in the future algo-culture will play a vital role in meeting the food needs of our protein-hungry world. Algo-culture systems can produce up to 15,000 kilograms of protein per acre per year. This is almost 20 times the per-acre yield of soybeans, which is the highest yield obtainable through conventional agriculture. Through algo-culture, in theory, a protein supply adequate for the entire planet could be produced in an area the size of the state of Maine.

Acres Required to Produce Protein Levels Equivalent to One Acre of Algae Production

Chlorella	1 acre
Corn	49.2
Hay	69.0
Wheat	95.4
Oats	108.0
Soybeans	20.9
Sorghum	75.0
Barley	994.0

Based on data from William J. Oswald and Clarence G. Golueke, Large-Scale Production of Algae, in Mateles and Tannenbaum (eds), *Single-Cell Protein*, The MIT Press, 1968, p.294.

Drs. Oswald and Golueke report, "In studies with our large-scale pilot plant at Richmond, California we demonstrated that with proper cultivation, at least 20 tons (dry wt.) of algae having a protein concentration of 50 per cent will be produced per acre of pond per year. In terms of yield of digestible protein on an areal basis, this yield is 10 to 15 times greater than that of an acre of land planted with soybeans and 25 to 50 times that planted with corn. From the standpoint of food energy, our algal cultures have yielded dietary energy on an areal basis at rates 8 times as great as that of sugar beets, 22 times as great as great as that of corn, and 45 times as great as that of potatoes. It is therefore quite obvious that controlled microalgal culture is potentially a more productive use of land for protein than is conventional agriculture."[1]

Dr. Lee established that Chlorella could replace one-third of the protein supplied by eggs and up to two-thirds of the protein supplied by fish for adult humans, without impairment of nitrogen retention. In a study covering ten experimental periods of five days each, the lowest nitrogen digestion observed was 66% in diets consisting of algae alone. Higher rates of digestion, up to 75%, appeared when algae were combined with other protein. These findings seem to indicate that algae are metabolized more efficiently when consumed in small amounts or when combined with more digestible proteins.

Richest Source of Chlorophyll

Although Chlorella and other microalgae such as Spirulina are perhaps best known for their protein content, they contain many other nutrients that are more important to us who live in Western countries with relatively abundant sources of protein.

One particular characteristic of Chlorella has led some to call it "the supreme whole food supplement." This is its high concentration of chlorophyll, which is often as high as 7% of its total weight. Chlorella is by far the richest source of chlorophyll available for human nutrition today. Alfalfa, for instance, contains about 0.2% chlorophyll. Chlorella contains almost ten times more chlorophyll than Spirulina (0.76%), and more than most of the processed chlorophyll supplements available in health food stores which rarely contain more than 4-5%. Jeanne Rose, the well-known herbal practitioner and author, believes that chlorophyll is the most important component of microalgae.

Chlorophyll, as much as any existing biological substance, deserves the title of "nature's healer." Its effectiveness is recognized by many naturopathic healers, doctors, and research scientists; its remarkable benefits have been noted in professional journals such as *The American Journal of Surgery* and *The New England Journal of Medicine*. The first comprehensive report on the therapeutic uses of chlorophyll appeared in the July 1940 issue of *The American Journal of Surgery.* In this report, many distinguished doctors reported on cases that ranged from deep internal infections to skin disorders to advanced pyorrhea. All these disorders responded positively to treatment with chlorophyll.

The catalogue of chlorophyll's positive effects is astonishing in its range. Here are some of its benefits, as detailed by Dr. Bernard Jensen, respected naturopathic practitioner and author of *Health Magic Through Chlorophyll*:

Provides iron, builds red-blood count and improves anemia
Removes toxins, cleans and deodorizes bowel tissue
Purifies the liver and aids hepatitis
Heals sores, soothes inflamed tonsils, ulcers and painful hemorrhoids and piles
Feeds heart tissues and improves varicose veins
Regulates menstruation and improves milk production
Aids hemophilia, improves diabetes and asthma

Chlorophyll as a Blood Builder

Chlorophyll is literally the blood of plants. Its chemical structure closely resembles that of hemin, which combines with protein to form hemoglobin in the human bloodstream. Chlorophyll and hemin molecules are, in fact, almost identical in structure, the only difference being that the chlorophyll molecule is built [7] around an atom of magnesium whereas hemin is built around an atom of iron. For this reason, chlorophyll has proven useful in building up the red blood cell count in humans. It is the red blood cells that carry oxygen to the tissues, and a low red blood cell count results in anemia. Many naturopathic practitioners recommend high-chlorophyll foods as an important feature of diets designed to prevent or cure anemia.

A high red blood cell count is one of the prerequisites of good health, high energy, and immunity to disease. This is why the red blood cell count is one of the first things routinely tested by doctors. The chlorophyll provided by Chlorella can play an important role in maintaining and improving this vital aspect of health.

Chlorophyll Detoxifies

A dangerous problem facing humanity today is the rising level of radiation. We can control our exposure to many forms of pollution, such as chemicals in food or synthetic drugs, but there is little we can do to reduce the amount of radiation we receive. Certain foods, however, appear to have the ability We can control our exposure to many forms to eliminate radioactive materials from the body. Miso, a fermented soybean paste popular in the Orient, is one. Its beneficial effects were discovered by the Japanese after the first atomic bomb was dropped on Hiroshima. Later it was discovered that the active ingredient, which occurs naturally in Miso, is a compound called zybicolin.

Tests performed by the U.S. Army showed that chlorophyll-rich foods may also be effective in decreasing the effects of radiation. In one controlled study it was found that a chlorophyll-rich diet doubled the life span of animals exposed to fatal doses of radiation. Dr. Bernard Jensen states that chlorophyll can be used as an antidote to pesticides and can help eliminate drug deposits from the body.

Chlorophyll is a powerful cleanser and builder of the hemoglobin in the blood. It helps remove toxic materials from all internal organs, thereby allowing a natural healing process to take place. Even people in good health can also experience great benefits from raising their consumption of chlorophyll because its powerful action strengthens resistance to all kinds of disease.

As the world around us becomes more industrialized, and a greater proportion of the population is forced to live in congested urban areas, we need all the protection we can get against the many dangerous substances that occur in our food, in our drinking water, and in our air. Chlorella with its high chlorophyll content and detoxifying properties promises to offer all-round protection.

Chlorella is the most highly concentrated source of natural chlorophyll available; moreover, it is a whole vegetable food that contains many other protective, health-building vitamins and minerals. Chlorella enables us to take large amounts of chlorophyll without using chemically extracted products. For instance, one tablet of Chlorella contains as much chlorophyll as about 35 tablets of alfalfa, which is often taken as a chlorophyll supplement.

Chlorella Growth Factor (CGF)

Dr. Fujimaki of the People's Scientific Research Center at Koganei in Tokyo discovered a physiologically activating substance which accelerates the growth and development of new cells in organisms in Chlorella called the Chlorella Growth Factor.

This physiologically activating substance is unique and is produced only in the process of rapid multiplication with photosynthesis. The structure of CGF is quite complex. It is composed of a nucleotide-peptide containing sulphur, polysaccharides and other substances.

Its molecular weight is in the range of 3-13x10 to the 3rd power and it can be dialyzed with a cellophane membrane. The main sugar element of the nucleotide is glucose, but it also contains munnose, rhamnose, arabinose, galactose and zylose. The amino acid structure of the peptide includes glytamic acid, aspartic acid, alanine, serine, glycine and prolamine. The greatest absorption takes place at ultra violet wave lengths of 260 mu. The least takes place at 240 mu.

Nucleic acids are important to the survival of all things as they are contained in all cells. It is this nucleic acid which controls reproduction of new cells, cell division, cell growth and the production of energy. Polysaccharides are effective in stimulating the immune system.

CGF is the most valuable substance contained in Chlorella and only in Chlorella can this substance be found. This is what makes Chlorella the most beneficial and popular health food today in Japan.

Research is being done, not only in Japan, but all over the world to further understand the positive effects that CGF has on the functioning of the human body.

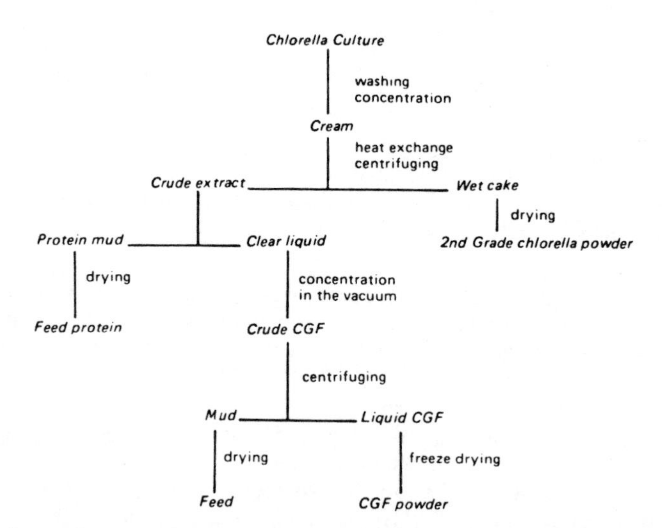

Flow sheet for CGF production. Reprinted by permission from: Production and Development of Chlorella and Spirulina in Taiwan, Pinnan Soong in Shelef and Soeder (eds), *Algae Biomass*, Elsevier/North-Holland Biomedical Press, 1980, p. 103.

Vitamins and Minerals

Chlorella is a rich, all-vegetable source of many nutritive substances. One of the most important of these is pro-vitamin-A, the form of Vitamin A which is most easily digested and which cannot build up to toxic levels in the body.

After protein deficiency, lack of Vitamin A is the most common deficiency on the planet. A United States Department of Agriculture survey in 1968 revealed that even in America, 25% of the population were deficient in this vitamin. Today an adequate source of Vitamin A is particularly important because this substance is depleted in the body by a number of environmental factors. For example, its assimilation and storage are inhibited by low-level exposure to pesticides.

Vitamin A plays several significant roles in the functioning of the body. It is essential, for example, for the formation of visual purple in the retina of the eye. Visual purple is the substance that enables us to see at night. It is also important for healthy lungs. MIT scientists, at a symposium on lung biochemistry in 1970, reported that Vitamin A helps protect the lungs from the two major components of air pollution, ozone and nitrogen dioxide. Strong indications have been found that Vitamin A prevents the formation of pre-cancerous cells. Dr. Saffioti, a director of the National Cancer Institute, reported at the Ninth International Cancer Congress that Vitamin A can help prevent lung cancer. Additionally, Vitamin A is important to the growth of infants and children. Consequently, it is important that there be an adequate supply of it in the milk of nursing mothers.

Just one tablespoon of Chlorella provides about 200% of the minimum daily requirement of Vitamin A in its safest form, pro-vitamin-A.

A Perfect Food?

Chlorella is remarkable in that it contains nearly all the nutritional elements required for a healthy life. For example, Chlorella contains the following basic nutritional ingredients: Nutrition Information (per 100g)

Calories	400-460k cal
Protein	55-65g
Carbohydrate	20-25g
Fat	5-15g
Vitamin A	55, 500 IU

Analysis from Sun Chlorella Co. Ltd., Kyoto, Japan

Another common deficiency is folic acid. Chlorella contains almost twice as much folic acid as beef liver, which is usually thought of as the richest source. During pregnancy inadequate amounts of folic acid can lead to miscarriage. Many women with histories of miscarriages have had normal deliveries after folic acid deficiencies were corrected. The November 1970 issue of the *Journal of the American Medical Association* reported that folic acid deficiencies are more common among women who use birth control pills. These drugs appear to inhibit absorption of folic acid.

Niacin has been used in massive doses to treat schizophrenia. Smaller doses are calming and contribute to emotional balance. The Miami Heart Institute has reported success in using niacin as a nutritional supplement in a program designed for people who have suffered heart attacks. Chlorella is one of the richest natural sources of niacin. One tablespoon provides 100% of the daily adult requirement.

Vitamin B12 has the most complex chemical structure of all the vitamins and is the only vitamin which has not yet been synthesized. Adequate levels of B12 promote feelings of well-being, good appetite, and a high energy level. It helps prevent deterioration of the cells.

Muscle meats and beef liver are the richest traditional sources of B12 liver, but it is found in much smaller quantities in milk and cheese. The threat of B12 deficiency and the possibility of pernicious anemia in extreme cases is a common argument against vegetarian diets. However, the availability of microalgae products are great news for vegetarians. Chlorella contains more B12 than beef liver, and is an excellent, clean source of this important "feel-good" vitamin. One tablespoon of Chlorella provides 333% of the daily adult requirement of Vitamin B12.

Calcium, the most plentiful mineral in the body, is the basic substance of bones, teeth and nails, and is essential for the correct functioning of the nervous system. Hyper-nervousness and difficulty in sleeping are often related to calcium deficiencies. Since all the calcium in the body is replaced every six years, and may not be replaced adequately, such deficiencies are quite common. Although milk is high in calcium, the process of pasteurization creates changes in its chemical structure which make the calcium hard to digest. Many naturopathic practitioners believe that this hard-to-digest calcium from milk builds up in the joints and contributes to arthritis. Chlorella contains significant amounts of calcium. One tablespoon provides 120% of the daily adult requirement.

Iron is essential to the formation of the red blood cells that carry oxygen to all parts of the body and is naturally associated with vitality and a healthy-looking complexion. Chlorella is an excellent source of iron. This fact, in addition to the high chlorophyll content of Chlorella, makes it a potent tonic for the bloodstream. Low energy levels and feelings of depression can result from an inadequate intake of iron. Extreme iron deficiency leads to anemia. Iron is particularly important for women, who need one and one-half times as much as do men. One tablespoon of Chlorella provides 320% of the adult daily requirement of iron.

Chlorella is also a good source of zinc, which is essential to the proper functioning of the brain, and is frequently recommended for those who are recuperating from illnesses or operations affecting the brain.

Chlorella is more than a collection of important vitamins and minerals. Being a whole plant, it contains all the complex chemical compounds necessary to the protoplasm of life. Usually we eat the seed, the root, the leaf, or the fruit of a plant. When eating Chlorella or other microalgae we eat the whole plant, getting the elements necessary for the plant's reproduction, its growth, and the maintenance of its life cycle.

The Digestibility of Chlorella

The nutrition contained in the Chlorella cell is highly concentrated, but the durable cell without special processing wall makes Chlorella hard to digest. Most of the digestibility problems reported have arisen from sun-dried Chlorella. Over a period of more than twenty years of Chlorella production and use in Japan, many advances have been made in processing the cells to improve digestibility.

Drs. Tannenbaum and Miller demonstrated that the nutritional value and the degree of utilization of unicellular protein can be enhanced by processing it. Protein digestibility, biological value, and net protein utilization (NPU) were improved when the cells were ruptured by passing them through a homogenizer. Digestibility was increased from 55.6% to 67.3%, biological value from 62.3% to 70.0%, and NPU from 34.5% to 47.1%. Dr. Labuza discovered that spray drying resulted in less nutritive and functional change than other methods, and therefore recommended this process for commercial use.

The spray-fed, single-drum dryers, however, that are generally used to process the algae, are expensive to buy and to operate. Efforts are currently in progress to find simpler alternatives. According to Dr. Venkataraman, one breakthrough is drum drying. In this process the cell walls explode when the algae is exposed to 120 degrees Fahrenheit for about ten seconds.

Dr. Enebo recently published a report on methods of breaking the cell wall and releasing the protein from the microalgae. Among the methods he describes are the use of mechanical breakdown using the stomach juice of the snail *Helix pomatia*. Researchers Mitsuda, Yasumoto, and Nakamura claim that urea soaking is an excellent method of processing algae to release nutrients. Researchers Cook, Lau, and Bailey have demonstrated that steam cooking for about 20 minutes improves digestibility.

The Sun Chlorella Company of Kyoto, Japan has developed a process for breaking down the cell wall. The process involves a physical disintegration of the wall without use of chemicals, acids or other impurities. They have patented the process.

Clearly, further research into the processing of algae should be a high priority. All that is necessary to make the high protein available is a low cost processing technology.

Chlorella and Children

Young children seem to be naturally attracted to Chlorella. Once they've tried it, they often insist on eating it every day. This may be because children still have the inborn instinct that impels them to eat what is good for them. Chlorella is packed with the kind of nutrition that growing children need, such as protein, Vitamin A, calcium, iron, and Vitamin B12, in a natural, unrefined form. In addition, many children are more susceptible than adults to colds, flus, and excess mucus. The abundant chlorophyll in Chlorella helps in the prevention and cure of all these disorders. Jeanne Rose suggests that a child's health, energy level, and resistance to colds are all improved by a diet that includes Chlorella and Vitamin C.

The effects of Chlorella on children were investigated in a program at the Medical Division of Nagasaki University in Japan. One group of children was given a Chlorella supplement with their daily diet. The control group received the same diet without Chlorella. The researchers reported that after six weeks the children who received Chlorella showed a measurable increase over the control group in muscular strength, measured by hand-grab and back-muscle tests. They also caught fewer colds and flus than did the control group during the course of the experiment.

Because Chlorella is a whole food, there is little danger that children will take too much of it, as they might with many synthetic or highly processed vitamin supplements. It is difficult to determine the sources or processing methods used to produce most vitamins available today. Chlorella, on the other hand, is one of the few potent food supplements for children that is totally natural.

How to Use Chlorella as a Daily Food Supplement

Chlorella provides many unique benefits unavailable in processed multi-vitamin. Most notably, it contains elements, such as chlorophyll, that are not included in any multi-vitamin currently on the market. Chlorella's high concentration of nutrition is not produced by any process of refinement or chemical extraction; it occurs naturally in the Chlorella cell.

Chlorella powder can be mixed directly with water or with fruit or vegetable juices. And it can be used in a number of cooked dishes. For example, a spoonful or two can be added to soups, salad dressings, and dips. Some people keep the powder in a shaker next to their salt and pepper so that they can easily sprinkle it into cooking food. When a small amount of Chlorella is added to a dip mix, the dip takes on a pleasant pale green color similar to that of a creamy Guacamole. Chlorella is especially beneficial when used with wheat products, such as noodles, because its high lysine content compensates for the low level of lysine in wheat.

Chlorella powder is often pressed into tablets, without the use of any binders or fillers. Tablets provide a convenient way to take Chlorella with, or instead of, other vitamin products. The tablet form is especially useful for traveling, because it can provide substantial nutrition wheN it is difficult to find the kinds of healthy foods prepared at home.

Recommended Dosage

Although you may take as many Chlorella tablets per day as you like, the typical user in Japan reports that they use:

—For general prevention and maintenance: approximately 15-20 tablets (3-4 grams) per day.

—When actual symptoms become apparent indicating the necessity of Chlorella treatment: approximately 25-30 tablets (5-6 grams) per day.

From: Japan Chlorella Treatment Center, Kyoto, Japan.

Taken as a daily food supplement, Chlorella produces noticeable health benefits. It provides the highest level of chlorophyll available in any form. Because chlorophyll is such a powerful detoxifier and so greatly aids the entire digestive process, a high-chlorophyll diet increases our ability to absorb all the other nutrients in Chlorella and in other foods. Indeed, thousands of people who have used Chlorella regularly report that it is the one food supplement that really makes a difference in the way they feel.

Possible Temporary Reactions

Chlorella may be taken alone or in conjunction with medications. No negative side-effects have been found in using Chlorella.

The following initial reactions may be noticed:

—Intestinal gases may be released due to rejuvenation of the peristaltic action of the intestines. This will cease as the intestines become cleansed.

—Irregularity of bowel movement, nausea or slight fever may be noticed in a small number of people. This symptom usually disappears within 2-3 days, very occasionally taking up to ten. Again, this is merely an indication that the Chlorella is actively working. These reactions are most prevelant in those persons who need the Chlorella the most.

—Allergy sufferers sometimes break out in pimples, rashes, boils or eczema, in some cases accompanied by itching. This means that the drive to regain homeostasis is being accelerated and the body is actively working to expel toxins.

—Bowel movements may become greenish in color. This simply means that excess chlorophyll is being expelled.

The above reactions should not be taken as side-effects but as favorable reactions which appear as part of the body's process while taking Chlorella. These indications show that the Chlorella is working.

FROM The Japan Chlorella Treatment Center.

Cooking with Chlorella

The most striking thing about Chlorella is its color. A very small amount will turn anything containing it green. It may be mixed with foods of similar colors, such as green pasta; or its color may be masked by darker colors, such as that of brownies, or it can be used in recipes that take advantage of its color, such as Aquamole™, a guacamole-style chlorella-tofu dip.

Chlorella's taste is distinctive. To get accustomed to it, you may wish to start by using it lightly — just a little each day. In the course of developing a taste, you can mix it with strongly-flavored foods such as onions, garlic, and spices. You will find that it has a rich aftertaste.

If the Chlorella has been processed, as is the case with virtually all commercially available Chlorella, you will receive the maximum value of its vitamins by cooking it lightly or not at all, or serving it in a drink or as a dressing. (Should you have any unprocessed Chlorella, you can break down its cell walls by steaming it for 20 minutes.) Blend or sprinkle Chlorella into well-cooked dishes just before serving. It is important to store it in air-tight and lightproof containers at cool temperatures. Remember, you get the best nutritive value from the protein in Chlorella if you eat it in combination with other proteins.

How you use Chlorella is limited only by your imagination and taste. If you've never tried cooking with microalgae try the following delicious recipes which will help you get started. Many of the recipes are based on those in Earthrise™ Spirulina promotional brochure. Try them with Spirulina, another nutricious microalgae available at your local health food store.

Recipes

Emerald Nectar

8 oz. apple juice
1-2 tablespoon lemon juice
½ tablespoon Chlorella
1-3 dashes cinnamon

Mix in blender for ten seconds. Drink the sweet green-apple foam as fresh as possible.

Notes:

Emerald Veggie

8 oz. tomato and/or carrot juice
1 teaspoon soy sauce
1 tablespoon lemon juice
1 tablespoon Chlorella
Dash of cayenne

Blend until smooth and enjoy.

Notes:

Emerald Pesto

1 bunch fresh basil
1 tablespoon Chlorella
¼-½ head garlic (more if you dare)
¼ cup olive oil
½ cup Parmesan cheese
1 tablespoon lemon juice
¼ cup pine nuts
1 lb. pasta

Combine basil, Chlorella, garlic and olive oil in blender and puree. Lightly toast pine nuts in olive oil. Combine emerald puree, cheese, lemon juice and pine nuts. Stir into hot pasta. The heat from the pasta will cook the garlic lightly, removing its bite. Delicious!

Notes:

Emerald Miso-Mushroom Soup

For Two

2 tablespoon butter
½ cup diced mushrooms
½ thinly slice onion
1-¾ cup bouillon/soup stock
7 tablespoon miso (mix varieties)
1 tablespoon Chlorella
2 tablespoon Parmesan cheese
1 teaspoon Worcestershire sauce
Salt and pepper
1 teaspoon lemon juice
1 tablespoon minced parsley

In a skillet, melt the butter; add mushrooms and onion and saute until lightly brown (about 4 minutes). Add 1-½ cups soup stock and bring to boil. In a bowl, add miso and Chlorella to remaining soup stock and mix until dissolved. Mix into hot soup. Add cheese, Worcestershire sauce, salt and pepper and bring to boil. Remove from heat and stir in lemon juice and garnish with parsley.

Notes:

Emerald Clam Dip

1 small package cream cheese
1-½ teaspoon lemon juice
½ teaspoon soy sauce
1 small can minced clams
1 tablespoon butter
1 tablespoon Chlorella
1-2 large garlic cloves
Pepper or tabasco sauce

Drain the clams and mince the garlic cloves and blend into the cream cheese. Add lemon juice, soy sauce, melted butter. Season to taste with pepper or tabasco. Serve with raw fresh vegetables or chips.

Notes:

Aquamole™

1 minced garlic clove
2 tablespoon Chlorella
2 tablespoon yogurt or mayonnaise
1 cube chicken bouillon (crushed)
½ teaspoon sugar or honey
2 teaspoons soy sauce
Ground pepper
½ cup tofu

Optional: 3-4 tablespoon ground toasted sesame seeds, cayenne or tabasco

Combine all ingredients and puree. This is delicious in tacos or on crackers.

Notes:

Emerald Candy

1 cup unsalted nut butter (peanut, almond or cashew)
1 cup honey or maple syrup
1 cup carob powder
½ cup toasted sesame seeds
½ cup Chlorella

Melt nut butter and honey over low flame. Stir frequently to prevent burning. Stir in carob, sesame seeds and Chlorella. Blend thoroughly until mixture begins to shine. Pour into pan and refrigerate at least 2 hours. Cut into squares. Great for parties and snacks.

Notes:

An Aid in Weight Control

Being overweight is one of the major health problems facing America today. Fortunately, it is one of the easiest to remedy. When the functions of the body are balanced, and the processes of assimilation and elimination are working correctly, the body will naturally revert to its ideal weight. Moreover, the desire for food will manifest itself in harmony with the maintenance of this balance.

Unfortunately, this natural way of functioning is often upset by the processed foods that constitute the bulk of the American diet today. The nutritionally empty calories of sugar, fats, white flour, and denatured food products do not provide the nutrition we need for health. In many people this lack of nutrition results in cravings for more food, which can lead to overeating. The effects of overeating spiral upward, because a diet heavy in refined foods tends to create a sluggish digestion, constipation, and a build-up of toxic materials in the digestive tract. In turn, this process inhibits proper assimilation of vitamins and other vital elements in food, which leads to an even greater desire to overeat. A hard-to-break cycle is thus established, one that most dieters quickly recognize.

Chlorella tables are not "diet pills" that inhibit appetite. They work to restore the natural balance of the body so that a real and permanent change can be made in overall health and weight. Thus Chlorella can be incorporated as a healthy and effective addition to any well-balanced weight loss program.

Real progress can be made toward permanent weight loss when the intestinal tract is cleansed and functioning properly, when the excess material that caused the congestion has been eliminated from the intestines and they can once more function normally.

Drs. Saito and Okanao found that Chlorella actually stimulates the peristaltic action of the intestines, thus promoting a speedy, healthy digestive process. Interestingly, Dr. Bernard Jensen reports that chlorophyll has this effect. It may be, then, that the high chlorophyll content of Chlorella is responsible for this stimulation. The chlorophyll in Chlorella feeds the friendly bacteria in the stomach, and it is well known that chlorophyll is a good neutralizer of stomach acids. A grandmother from San Francisco writes:

> I am 73 years old and have always had difficulty with my digestion, and have also suffered from severe constipation. Now that I take Chlorella regularly, I no longer experience any of these problems. I also have greatly increased energy and have lost several pounds of excess weight.

The fact that Chlorella stimulates peristalsis in the intestines is undoubtedly one reason it is such an effective addition to any weight-loss program. Sluggish digestion is often one of the contributing causes of overweight. It is hard to make real progress toward a permanently normal weight until this problem is cleared up. When the digestive system is functioning properly, the body can effectively eliminate the excess material that contributes to overweight. Chlorella's beneficial effects on the peristaltic action of the intestines will immediately assist in the cleansing action vital to this process. In addition, the high levels of easily assimilated nutrition in Chlorella provide the real food needed by the body, thus reducing the cravings that lead to "eating binges."

Yamagishi, head of the Clinic Hospital of Tokyo, studied infants who were unable to digest milk formula and who even developed allergic reactions to it. When a formula made with Chlorella was used, the infants digested it with no problems. If even these babies can assimilate Chlorella, anyone can!

Through scientific testing — and through its use over the past 25 years by millions of people of all ages — it has been proven that the super-nutrition in Chlorella can be assimilated by even the most delicate systems. In fact, Chlorella has positive effects that actually improve the entire digestive process. The stomach is the "power plant" of the body, and the benefits it receives from the regular use of Chlorella are reflected in higher levels of health and energy.

Many people have experienced great success in using Chlorella as part of a weight-loss program. An award-winning television producer wrote:

> I was about a hundred pounds overweight and, though I had tried numerous diets, I was unable to lose this excess weight. However, when I began to include one-fifth of an ounce of Chlorella in my natural diet, I noted an immediate improvement. After one year I had lost all of the excess one hundred pounds. My mother had a similar problem and also achieved excellent results in weight loss with a combination of Chlorella and natural foods.

Because Chlorella works to restore the natural balance of the body, it is also very good for people who need to build appetite and muscle. For those who are weak or debilitated, Chlorella is an excellent food supplement for restoring the energy necessary to rebuild health and strength.

How to Use Chlorella With Your Weight-Loss Program

Chlorella can be used effectively with any well-balanced weight-loss program. The beneficial effects of its cleansing properties often bring results without any strict changes in the daily diet. For the best results, four or five tablets should be taken three times a day before meals. If powdered Chlorella is used, about one teaspoon can be stirred into a glass of water or juice.

Because Chlorella is a whole food, it does not suppress the appetite. Rather it satisfies appetite while providing the energy and well-being that result from good nutrition. Many people who use Chlorella discover that they simply lose their desire for unnecessary meals and snacks, and do not even miss them!

There are hundreds of slimming diets, and almost as many theories as to how to lose weight. The experience of most dieters indicates that strict diets do not usually work permanently. Once the dieting period is over, the old habits, which caused the overweight in the first place, return. The real necessity for long-term weight control is to establish new, more enjoyable, healthier eating habits.

A good daily diet should contain large proportions of whole, unprocessed foods such as whole-grain breads, salads, fresh fruit and juices. These will naturally encourage the dieter to eat less sugar, sweet baked goods, and heavy foods such as fats and meats. As we discover the wide range and the delicious flavors of the many foods Nature has to offer, our cravings for "junk food" naturally diminish.

A diet such as this tends to re-establish the natural balance of the body, so that severe weight problems can be permanently relieved. Chlorella is a powerful tool that works with other natural foods to accelerate this process.

Chlorella and Hunger Control

Much has been written recently about the ability of the amino acid phenylalanine to suppress the appetite center of the brain. A recent article in a major national magazine actually connected the high phenylalanine content of Spirulina with its ability to aid in the loss of weight. Chlorella contains a slightly higher percentage of phenylalanine in its protein than does Spirulina. However, it is premature to make definite claims that this amino acid is effective in weight control, because research into the question is still in progress.

Whether or not research proves these claims to be true, however, it would be a mistake to focus on this "miracle pill" aspect of Chlorella in relation to weight-loss programs. Although the manufacturers of diet pills would have us believe otherwise, there really is no such thing as a miracle diet pill. Chlorella performs miracles, by aiding our bodies, the real miracle workers, to tune into the natural energies and instincts for health that lie within each one of us.

The Chlorella Slimming Diet

Each of us has an individual set of nutritional requirements: we all need different quantities and kinds of food. These requirements arise from factors such as lifestyle, metabolism, and personal preferences. For this reason, it would be impossible to impose one diet successfully on everyone. Nevertheless, some guidelines can be helpful. Here is the outline of a weight-loss and health-building diet which is satisfying, effective, and flexible enough for almost anyone to follow.

Breakfast

Start the day with one teaspoon of Chlorella dissolved in fresh-squeezed fruit juice (such as orange, grapefruit, or pineapple), or in fresh lemonade made from lemon juice and water, sweetened with a little maple syrup or honey to taste. If you like, you may substitute five Chlorella tablets taken with herbal tea. Follow this with a meal of whole grain toast, fruit salad with yogurt, a portion of hot or cold whole-grain cereal, or any desired combination of these dishes.

Midmorning

If you feel hungry at this time, eat some fresh fruit with tea or coffee.

Lunch

Again, begin your meal with a teaspoon of Chlorella dissolved in fruit juice, or with five Chlorella tablets. Follow with a satisfying meal made up of one or more of the following: salad, steamed vegetables, soup, sandwich on whole-grain bread with plenty of greens and sprouts, baked potato, fresh fruit.

Dinner

Once more, start with one teaspoon or five tablets of Chlorella. Satisfy your hunger with a meal that includes a good salad as well as any vegetable dishes (for example, beans, lentils, or baked potato) that appeal to you.

As you can see, this is not the kind of strict diet usually recommended for weight loss, in which calories are religiously counted and painful deprivation is demanded. Rather it emphasizes the importance of eating a satisfying amount of the right kinds of light, healthful, and unprocessed foods. Dairy foods, such as cheese and milk, can be eaten in moderation, but it is best to avoid meat and eggs when trying to lose weight. These foods are high in fat, and tend to congest the system at a time when the dieter is seeking to correct the imbalances caused by sluggish elimination. The Chlorella Slimming Diet is a new concept in weight loss. It has been followed successfully by many people who were unable to make progress with any other method.

Chlorella and Spirulina: How Do They Compare?

In the future, we will be able to choose among many kinds of microalgae foods, each with a different nutritional profile and many different applications in food preparation. At present, the choices are limited basically to Chlorella and Spirulina. (Dunaniella, another unicellular green alga, is gaining in popularity.) In many areas, these two popular microalgae are similar, although there are a number of important differences. Both contain between 55% and 65% protein. Measurements vary from one batch to another because nutrients in the algae are influenced greatly by growth conditions. In general, Spirulina tends to have a slighly higher percentage of protein, but the difference is so small as to be of little significance. For instance, the Chlorella imported into the U.S. contains slightly over 60% protein; most of the Spirulina available at present comes from Lake Texcoco in Mexico, where the growing conditions produce considerable variations in the protein level, between 60% and 68%.

Difference Between Chlorella and Spirulina

Spirulina is a multi-cell spiral shaped plant, and is completely different from the round single cell of Chlorella. Although Spirulina and Chlorella may look similar, they are scientifically different. They belong to different systems, different class and different order.

Chlorella belongs to the Chlorophyceae class and Chlorococales order, whereas Spirulina belongs to the Cyanophyceae class and Nostocales order. Chlorella is single-celled while Spirulina is multi-celled. Chlorella has a nucleus and measures three to eight microns in size, whereas Spirulina is a hundred times larger than Chlorella and has no true nucleus. Spirulina is herical while Chlorella is spherical.

The main pigmentations produced by photosynthesis differ somewhat. Chlorella produces chlorophyll a, b and B-Carotene, whereas Spirulina produces chlorophyll a, b, B-Carotene and phycocyanin. They also differ in structure, Spirulina has neither chloroplasts nor a nuclear membrane.

Chlorella provides twelve times more iron than Spirulina, five times the chlorophyll and three times more calcium.

It has been stated that the high phenylalanine content of Spirulina suppresses the appetite center of the brain and aids in weight loss. Chlorella also contains a high percentage of phenylalanine to aid in weight loss. However, definite claims that this amino acid is actually effective in weight control are premature, as research is still in progress.

Protein Content of Chlorella and Spirulina

	Chlorella	Spirulina
% Protein	57%	60%
Amino Acids:		
Lys	6.09%	4.59%
Trp	1.27%	1.40%
Thr	3.25%	4.56%
Met	1.55%	1.37%
His	1.55%	1.77%
Val	4.58%	6.49%
Ile	4.00%	6.03%
Leu	6.00%	8.02%
Phe	4.03%	4.97%
Arg	6.30%	6.50%

Based on data from Park Waldrup, "Microorganism as Feed and Food Protein," in Altschul and Wilcke (eds), *New Protein Foods,* Volume 4, Part B: Animal Protein Supplies, Academic Press, New York: 1981, p.244

Spirulina contains more Vitamin B12 than Chlorella, although the amount of B12 in Chlorella is considerable. Again, large variations are observed from one batch to another. According to one large importer of Spirulina, who has had a number of analyses performed by independent laboratories, the amounts of this vitamin found in different batches varied by as much as 50%.

Comparison of Chlorella and Spirulina

	Chlorella	Spirulina
Chlorophyll	7.2%	.76%
Vitamin B12	1.02 mg/kg	2.00 mg/kg
Niacin	240.00 mg/kg	118.00 mg/kg
Calcium	3450.00 mg/kg	1315.00 mg/kg
Iron	570.00 mg/kg	528.00 mg/kg
Zinc	39.00 mg/kg	46.50 mg/kg

Nutritional profiles are very similar, except in the case of chlorophyll. Chlorella contains almost ten times as much chlorophyll as Spirulina. By comparison with other common sources of chlorophyll, Spirulina contains large amounts (for example, Spirulina contains 0.76%, alfalfa contains 0.2%). Chlorella, however, is a green alga, as opposed to a blue-green one, and contains up to 7% chlorophyll.

In combination with this high level of chlorophyll, Chlorella's high iron content makes it an unusually effective builder of red blood cells in humans. The red blood cell count is one of the primary factors on which health, resistance to infection, and the circulation of oxygen to the muscles and brain.

It is important to consider the differences in the methods of cultivation used to produce Chlorella and Spirulina. Although the Spirulina from Lake Texcoco is heat-sterilized to eliminate live bacterial contamination, it is not possible when harvesting a wild crop of this kind to exclude all other foreign organisms or materials.

For example, scientists Becker and Venkataraman, in their report on the results of their Indo-German pilot plant, caution that "algae are able to accumulate high amounts of substances from the medium which may lead to harmful side-effects in humans after consuming [them]. Most attention in this respect is given to the amount of heavy metals. . . found in algae. Contamination sources are: the water, the fertilizer, or emissions from industrial waste gases. Besides the fact that varying quantities of metals have been detected in algae grown under outdoor conditions, wrong and exaggerated estimations of this hazard have brought discredit upon potential utilization of microalgae."[2]

At present there is considerable controversy over the matter of contamination. Marketers of wild Spirulina claim that Spirulina pond contamination is minimal or nonexistent. However, several leading microbiologists specializing in algo-culture have cautioned that contamination can be a considerable problem in wild strains of microalgae harvested for commercial use.

Earthrise Farms in the Imperial Valley in southern California, the only commercial producers of Spirulina in the United States, have solved this problem. Their Spirulina is grown under controlled conditions and tested several times a day and before bottling by qualified scientists. Earthrise™ Spirulina has had no contamination problems. Similarly, Chlorella produced by the Sun Chlorella Company in Japan is cultivated in large modern facilities under sterile conditions. Consumption of products from these companies is safe.

A number of news features on television have reported gastrointestinal problems experienced by people who consumed wild Spirulina. All unicellular algae are known to cause some minor gastrointestinal disturbance during the first few days of use in large quantities. However, controlled studies show that this is part of the adjustment process and quickly passes. When Chlorella is integrated into one's diet in small quantities, it rarely causes intestinal discomfort.

Footnotes

[1] Oswald, W. J. and C. G. Golueke, Large-Scale Production of Algae, in Mateles and Tannenbaum (eds), *Single-Cell Protein,* The MIT Press, Cambridge, 1968, p.294-295

[2] Becker, E.W., and L. V. Venkataraman, Production of Algae in Pilot Plant Scale: Experiences of the Indo-German Product, in Shelef and Soeder (eds), *Algae Biomass,* North-Holland Biomedical Press, Elsevier, 1980, p.39

The Chlorella Story Bibliography

Altschul, Aaron M. and Harold L. Wilcke, *New Protein Foods,* V. 4, Part B, Academic Press, 1981

Bender, A. E., (ed), *Evaluation Of Novel Protein Products,* Perganon Press

Burlew, John S., (ed), *Algal Culture: From Laboratory To Pilot Plant,* Carnegie Institute, Washington D.C., 1953

Becker, W. E., and L. V. Venkataraman, Production and Processing of Algae in Pilot Plant Scale: Experiences of the Indo-German Project, in Shelef and Soeder (eds), *Algae Biomass,* Elsevier/North-Holland Biomedical Press, 1980

Calloway, D. H., The Place of SCP in Man's Diet, in Davis (ed), *Single Cell Protein*, Academic Press, 1974, p. 129-146

Enebo, L., *Evolution of Novel Protein Sources*, Pergamon Press, New York, 1968, p. 93-103

George, Uwe, *In The Deserts Of This Earth*, Harcourt Brace Jovanovich, 1977

Hills, Christopher and Hiroshi Nakamura, *Food From Sunlight*, University of the Trees, 1978

Jensen, Bernard, in Goldman (ed), *Health Magic Through Chlorophyll*

Jorgensen, J., and J. Convit, Cultivation of Complexes of Algae With Other Freshwater Microorganisms, in Burlew (ed), *Algae Culture: From Laboratory To Pilot Plant*, Carnegie Institute, Washington D.C., 1955, p. 190-196

Lee, S. K., H. M. Fox, C. Kies and R. Dam, The Supplementary Value of Algae in Human Diets, *Journal of Nutrition*, 1967, v. 92, p. 281-285

McDowal, Marion G., and Gilbert A. Leville, Feeding Experiments With Algae, U. S. Army Medical Research and Nutrition Laboratory, Fitzsimmons General Hospital, Denver, Co., 1963

Milner, Max, Nevin S. Scrimshaw, and Daniel I. C. Wang, *Protein Resources and Technology: Status and Research Needs*, MIT Press

Morimura, Yuji, and Nobuko Tamiya, Preliminary Experiments in the Use of Chlorella, *Food Technology*, 1954, v. 8, # 4, p. 179-182

Oswald, W. J., and C. G. Golueke, Large-Scale Production of Algae, in Mateles and Tannenbaum (eds), *Single-Cell Protein*, The MIT Press, Cambridge, 1968

Oswald, W. J., Advances in Environmental Control Studies With A Closed Ecological System, *American Biology Teacher*, 1963, Oct., v. 25, #6

Soong, Pinnan, Production and Development of Chlorella and Spirulina in Taiwan, in Shelef and Soeder (eds), *Algae Biomass*, Elsevier/North-Holland Biomedical Press, 1980, p. 97-121

Waldrup, Park W., Microorganisms As Feed and Food Protein, in Altschul and Wilcke (eds), *New Protein Foods*, Academic Press, 1981, v. 4, Part B, p. 228-249

The Amazing Alchemist

BEVERLY A. POTTER, PHD.

Microalgae Production

Most commercial production of microalgae involves Chlorella cultivation ponds in the Far East and Spirulina ponds in Mexico, California, and Israel. Until recently, progress was retarded by a number of technical and economic problems.

Production is expensive. Chlorella is produced primarily as a high-value health food which means that maintaining product quality is more important than reducing product cost. Health food prices can support markups of as much as 1000% from production costs of $10/kg. Consequently, there has been little incentive to reduce prices in this industry. Although the use of algae for animal feed holds great protential, early hopes for large-scale production of feed have not yet borne fruit. A market for algae in tropical fish foods exists but it is economically restricted, and has not attracted large commercial investments.

There are technical problems too. Besides the problem of breaking Chlorella's cell-wall, harvesting requires technically sophisticated and expensive centrifugation or chemical flocculation processes for separating the algae from the water. Another area of great potential is industrial chemicals but the low cost of oil-based products has suppressed their production. On the other hand, the rising price of oil, the search for natural product substitutes, and the growth of new agricultural technology are beginning to turn these problems around.

Cultivation

Open Air Ponding Systems

Cultivation of algae usually takes place in a shallow open air pond with channels that direct the flow of the water. Paddlewheels stir the water to keep the culture uniformly distributed, to even out the temperature, to prevent in-pond settling, and to distribute nutrients. Most of the expense is incurred in harvesting and drying the algae, purchasing high-quality chemical nutrients, reconditioning the water, and maintaining the integrity of the pond channels. Several hundred acres must be cultivated in order to support these costs.

Building concrete ponds is one of the main factors in the high cost of Chlorella production. Pinnan Soong, in Taiwan, developed a "red mud plastic," which is resistant to the sun's ultraviolet rays, as an inexpensive alternative to concrete ponds. In pilot tests the "red mud plastic" ponds have not corroded after five years of intense exposure to the elements.

Venkataraman, in India, constructed a modified tank system using brick and mortar which occupy an area of 128 square meters, and have a capacity of 12,000 liters. The system is divided into three sections: an outer shoulder, a middle tank with a sloping floor, and the central storage tank, which contains a sump pit. The tanks are connected to one another by a series of evenly spaced conduits (holes) in the walls of the middle tank and the central storage tank. A centrifugal pump, fitted on the bridge of the tank, moves the culture medium from the sump pit out to the shoulder. The algae medium then flows back to the center through the conduits between the tanks. The rate of flow is regulated by the size of the holes between the sections which reduces the energy required to agitate the cultures merely reduces production costs.

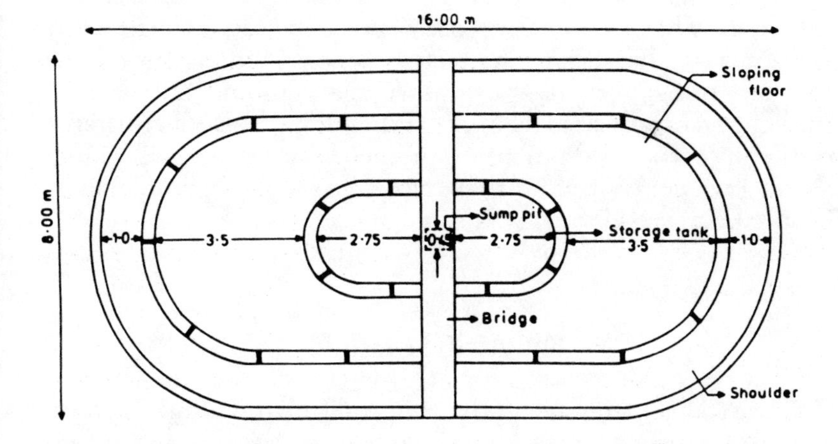

A schematic of the tank cultivation system built by Dr. Venkataramar, using gravity to stir this medium, resulting in significant energy savings (see text).

Closed System Fermenters

Chlorella may be cultivated in closed systems as well as in open ponds. Lack of sunlight to produce chlorophyll production is the major problem in closed fermenting systems. At the Weiwang Company, in Taiwan, Chlorella is produced in completely closed systems, from test tubes to small tanks to large fermenters. The concentrated stream of Chlorella is pumped from the large fermenter onto the roof of the factory where it then flows down through translucent plastic tubes while exposing the algae to sunlight and then returns to the fermenter. The sunlight transmitted into the medium through the plastic tubes produces a chlorophyll concentration that compares favorably with that of Chlorella grown in open ponds.

The Airlift Method

Genevieve Clement set up the first airlift algo-culture basin in southern France, using injected combustion gas. The device consists of a horizontal platform with a well at each end; the wells are each separated into two compartments that connect only at the bottom. Combustion gas is injected through the algae providing the nutritient carbon dioxide and causing circulation between the compartments which is necessary for efficient photosynthesis. Depending on velocity, mechanical stirring with paddle wheels, which is used in all other large-scale culture projects, uses considerable energy. By comparison, combustion gas uses less energy to circulate the algae mixture. However, Dr. Oswald, leading sanitary engineer, notes that paddlewheel stirring at 0.5ft/sec requires less energy than the airlift method.

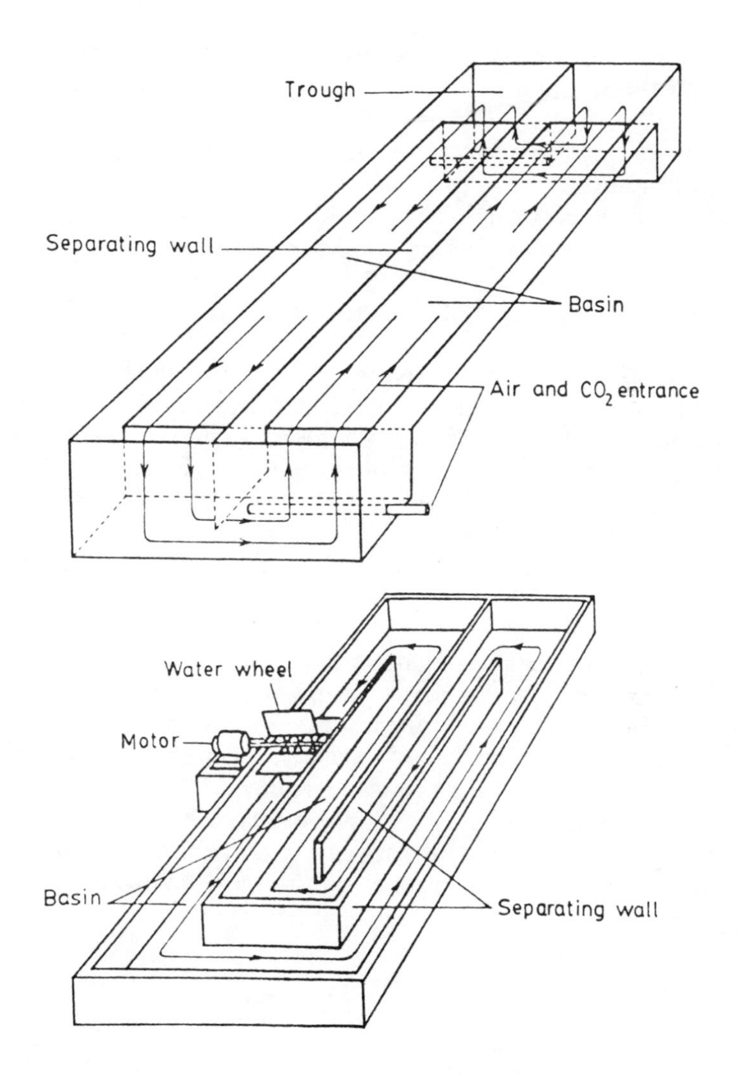

Schematic drawing of an airlife type culture pond (left) and a meandering raceway pond in which the flow is generated by means of a paddle wheel (right). Reprinted by permission from Pinnan Soong, Production and Development of Chlorella and Spirulina in Taiwan, in Shelef and Soeder (eds), *Algae Biomass*, Elsevier/North-Holland Biomedical Press, 1980, p.108.

Dr. Clement claims an average daily yield of 12 grams of dried matter per square meter (80 lbs./acre). This means that one hectare can produce between 40 and 45 tons of dried matter per year (16 to 18 metric tons per acre) with more than 65% protein content. The cost of algae produced by this ingenious method is low enough to attract further research.

Use of Flocculants

Separating the algae from the water can be difficult. Large filamented algae, such as Spirulina, are comparatively easy to remove from water. Unicellular algae such as Chlorella, on the other hand, are very small and therefore harder to harvest. Flocculants are often used in the separation process. A flocculant is a substance added to an algae culture which causes the algae to clump up or cluster, thereby making it easier and cheaper to separate algae from water.

A disadvantage of using flocculants is that they change the pH of the water, making it necessary to treat the water before reusing it or discharging it into the environment. Another problem with chemical flocculants is that they may be absorbed by the algae resulting in a high concentration of calcium or aluminum. The absorption problem may be overcome by using organic flocculants, or by using paddlewheel mixing techniques to promote "autoflocculation" by algae cultures that flocculate without additives.

How Chlorella is Manufactured

By The Sun Chlorella Company in Japan

The technology developed by the Sun Chlorella Company in Kyoto, Japan is designed with quality control at every stage of production to insure the purest and most consistently high quality Chlorella on the market.

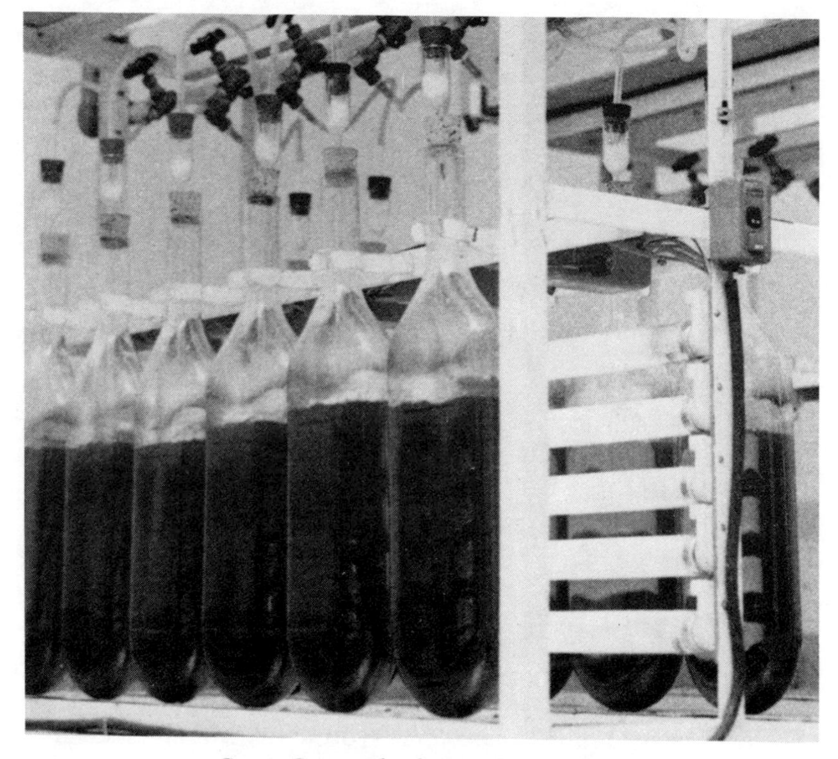

Step One: Flask Seed Culture

First the culture is grown indoors in flasks. Chlorella from an agar slant medium is implanted in small culture flasks containing nutrient solution. The culture is grown under fluorescent light and is treated with bubbling air and carbon dioxide.

Step Two: Factory Seed Culture

The factory seed culturing basin is a round shallow pond, made of concrete. A stirrer in the center supplies a continuous flow of nutrients to the Chlorella and stirs the culture medium. As the concentration of Chlorella increases, the depth of the culture pool is gradually increased.

Step Three: Main Culture

The main culture is grown in a concrete pool whose diameter is between 40 and 55 meters. Again, a mechanical stirrer is used in the center. Most Chlorella produced in Japan is manufactured in this way. This method is known as the open circulation culture system, to differentiate it from the closed culture system. In the closed system, culturing is accomplished with aeration, but without light.

Step Four: Purification and Condensing Process

The culture in the main pool is condensed to about one gram of raw Chlorella per liter of medium. At this stage, efficient removal of the water is crucial. It is not desirable to use a coagulant or flocculant, because the Chlorella is capable of absorbing large amounts of coagulant. The most efficient method is to condense the material by means of several steps of differential centrifugation and to wash it with a DeLaval-type centrifuge.

Step Five: Cell-Wall Disintegrating Process

Next, it is necessary to break down the durable cell walls characteristic of Chlorella so that its nutritive components will be available for digestion. The Sun Chlorella Company, of Kyoto (Japan), has a patented process for cell wall disruption.

Step Six: The Drying Process

Since raw Chlorella decomposes easily, it is necessary to keep it at a low temperature and to dry it as rapidly as possible. Spray-drying is the most common method of producing powdered Chlorella. The Chlorella concentrate is sprayed into hot air, which dries it instantaneously with little damage to its vital nutritive elements. Once dry, Chlorella is quite stable. If preserved properly, it will retain its nutritive value for several years.

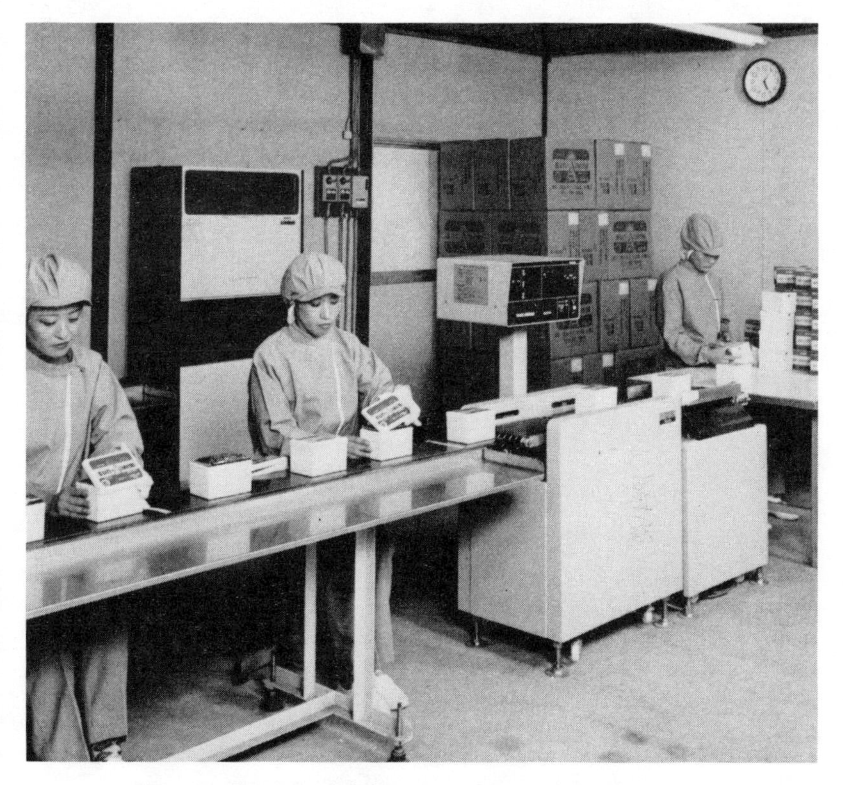

Step Seven: Tableting and Packing Process

After drying, the Chlorella powder is inspected for mutation or contamination. It is then pressed into tablets by a machine. No binding substances are added to the Chlorella. The tablets are packed in poly-propylene-coated aluminum foil, and finally in a poly-propylene case. Preceding photo-series courtesy of Sun Chlorella Company.

From Waste to Water

One of the most fascinating features of algae is its ability to transform toxic human, agricultural sewage into fresh water and food. Algae is truly the emerald alchemist. Algae promises to provide solutions to problems of waste removal and world hunger problems. Let's take a look.

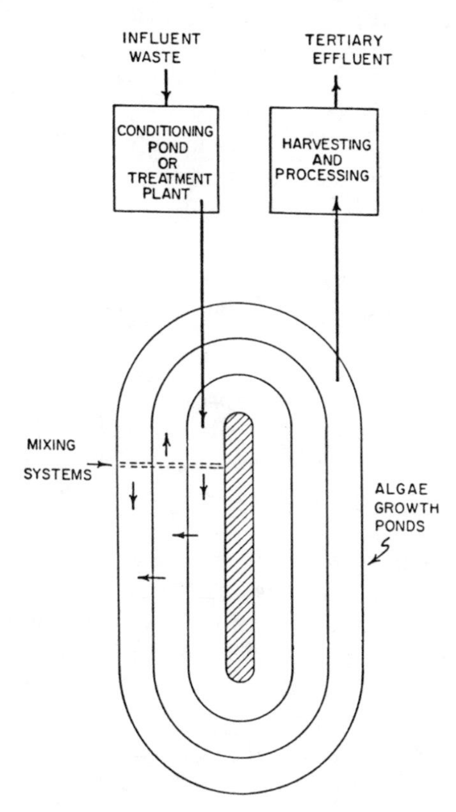

System design for large scale algae production. Diagram courtesy of Dr. William Oswald, University of California at Berkeley.

Domestic sewage is an excellent source of nutrients for algae. But first, the organic substances present in sewage must be broken down by bacteria in order to render the vital nutrients available to the algae. Dr. William Oswald, professor of sanitary engineering at the University of California, Berkeley, developed the concept of "integrated ponding systems" as a low-cost, energy-efficient approach to algal-bacterial interaction in waste treatment. Here's how it works. In a series of several ponds, bacteria oxidize organic wastes into nutrients usable by algae. Algae use sunlight energy to incorporate these nutrients during photosynthesis into their cells, giving off oxygen as a byproduct, which the bacteria then use in the oxidation of additional wastes. The outcome is that the elements of waste organic matter are converted into algal cells. Surprisingly, the weight of the algae yield usually exceeds the weight of the waste that has been converted. The difference is made up of water and carbon dioxide given off by the bacteria. Integrated ponding systems can be designed to treat domestic or agricultural wastes.

The Integrated Park-Pond

Integrated ponding systems may prove to be the ideal solution for treating wastes in outlying residential areas far from urban sewer lines. A basic pond design concept consists of a series of six ponds surrounded by a large park-like area integrated into the housing subdivision.

The sewage lagoon system is constructed as a treatment and disposal plant for domestic wastes at the end of a relatively long "outfall" or sewer system. Typically, they are designed as expandable so they can be enlarged to meet future population growth. By comparison, most traditional sewage treatment systems are not meant to be expanded. They are designed to serve a specific and fixed population in a definite area. Sewage lagoons, on the other hand, are made an integral part of the plan of the subdivision or trailer park it is to serve, in much the same way as septic tank-leaching field systems are integrated into the designs of single lots. The integrated park-pond concept has several advantages over septic systems, including avoidance of excessive lot size, reduction in future expenditures required if connection with a major sewer system becomes possible, and avoidance of discomfort, property devaluation, or health hazards resulting from the failure of individual septic tanks. In addition, the integrated system guarantees that the area that has been set aside around it remains a "green belt" or open space.

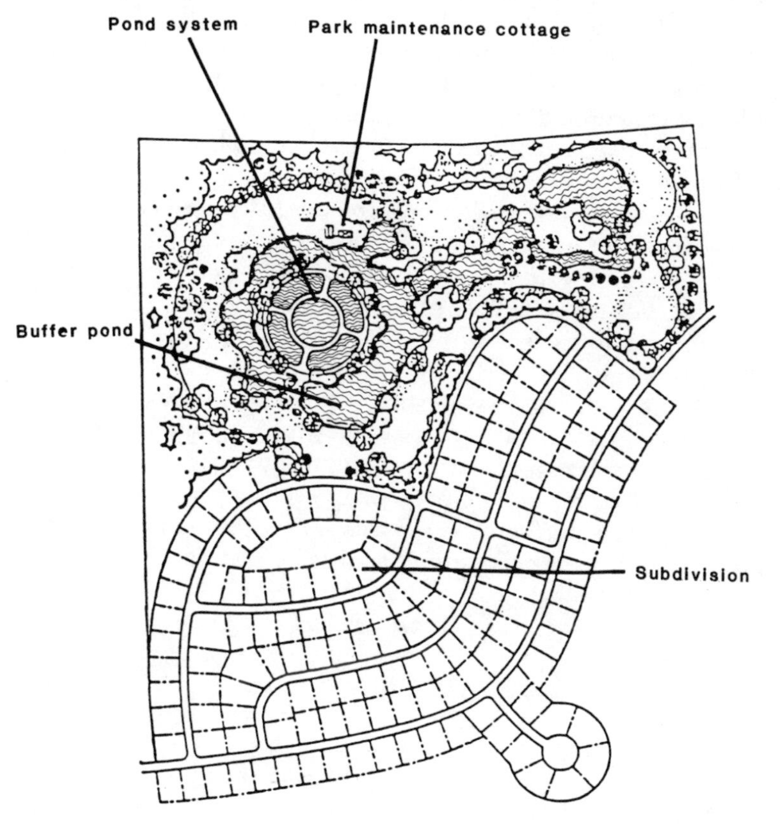

Residential sewage lagoon system. Courtesy of Dr. William Oswald, University of California at Berkeley.

How the Integrated System Works

Wastewater enters the centrally located primary pond at the bottom center through a vertical riser. Bacteria located at the pond bottom begin breaking down the waste and releasing nutrients essential for algal growth. Primary-treated wastewater is drawn from near the bottom of the first pond to insure that heat and grease are retained within this pond. The outflow from the first pond is transferred into the middle of one end of the second pond. It then moves in sequence to the third, fourth, and fifth ponds. Except in the transfer between Pond 1 and Pond 2, the surface water, which is the clearest, warmest water in the pond, is decanted and transferred to the next pond. Transfer to the sixth and last "buffer pond" is always indirect because this outermost pond may be used for water sports, such as sailing or fishing.

Amazing as it may seem, odors are virtually absent from the disposal area. In fact, the ponds give off less odor than a properly vented septic tank. Additionally, the ponds add to the beauty of the area as well because they often become havens for fish, ducks and other birds.

The major impediment to widespread adoption of sewage lagoons is acceptance by local authorities, who must approve developers' plans for waste disposal. Fortunately, authorities are becoming more knowledgeable and many now agree that properly designed ponding systems perform better than any other combination of treatment and disposal systems available at present for use in outlying areas.

From Waste to Food

Chlorella has tremendous potential as a feed for animals because, like other algae it efficiently converts sunlight and minerals into high-quality food. The durable cell wall is less problematic when using Chlorella as feed because grazing animals, such as livestock with multiple stomachs, can break down its cell walls and digest its nutrients just as they do when they eat grasses.

Prior to 1960 most animal feeding research with algae was performed on rats and chicks. The Grain Processing Corporation of Muscatine, Iowa produced about two million pounds of algae (grown on corn liquor) and fed it to chickens to enhance formation of yellow pigment in egg yolks. At about the same time, Dr. Combs fed chicks a diet containing Chlorella as a substitute for soybean meal. He found that when algae were substituted for 10% of the soybeans in the diet, a more efficient utilization of protein resulted and led to a significant improvement in growth. In 1957 Drs. Grau and Klein documented the nutritive value of algae grown on sewage and fed to baby chicks.

Dr. Leveille and others fed chicks and baby rats on diets containing various mixtures of Chlorella and other algae. They discovered that a mixture of several algal species was superior to any single species used alone. The Chlorella algae mixture was found to be especially effective when supplemented with methionine, an amino acid.

Dr. Lubitz, in 1963, experimented with freeze-dried Chlorella. He reported that digestibility was high and that the protein content was 55%. In 1973 Dr. Yamaguchi demonstrated that properly prepared Chlorella was an effective source of protein and that boiling the algae for 15 minutes improved digestibility and availability of protein.

Most raw powdered algae was not very palatable to livestock. This problem was almost entirely overcome by a University of California at Davis research group which included Drs. Harold Hintz, Hugh Heitman, James Meyers, Bill Wein and Dick Grau, when they made pellets from processed algae and steam-rolled barley. The cost of pelletizing algae is about ten dollars per ton of final feed or about one dollar per ton of algae.

These numerous early studies supported the conclusion that algae can be used as a high-quality feed for animals. At the same time, extensive systematic experimental work was being carried out with waste-grown algae. The use of algae to treat wastes opened up the intriguing possibility of multiple uses of Chlorella cultures — to break down waste and to create feed as a byproduct of the purification process. Between 1960 and 1966, Dr. Oswald cultivated algae in a one-million-liter production pilot pond in Richmond, California, and with the previously mentioned animal nutritionists at the University of California at Davis fed large animals with algae grown on waste products. The most interesting model developed by Dr. Oswald was the integrated feed lot, in which the manure of the livestock was used to feed Chlorella cultures, which were subsequently harvested and fed back to the livestock.

The Integrated Feed Lot

Ponding has been used to raise fish culture for hundreds of years and in waste disposal for over 70 years. Yet until Oswald's groundbreaking research over the last three decades, little scientific work had been done on combining a series of ponds into a waste to food alternative. In these systems algae are an integral tool in the management of key nutrients important to life, such as nitrogen and phosphorus. Algae production is a way to "fix" the nutrients after bacteria release them into water. When organic matter is decomposed by bacteria, carbon dioxide, ammonia, phosphate, and other substances are released. Typically, in classical agriculture and waste treatment, these nutrients are lost. The nitrogen and carbon dioxide go back into the atmosphere, and the phosphorus washes through the soil or is discarded with "sludge," the solid residue produced during sewage treatment. By contrast, in the integrated system, the nutrients are absorbed by the algae as soon as they are released.

Of course, most people are reluctant to use sewage-grown algae as food. Moreover, even though very few diseases can be transmitted between animals and humans, there is concern that domestic sewage may be hazardous when used as animal feed for domestic livestock. The real danger, however, is the buildup of heavy metals, herbicides, and pesticides found in sewage, rather than disease.

Human waste constitutes only a small fraction of the organic waste available. In the United States there is about five times as much industrial waste and 25 times as much agricultural organic waste as there is domestic sewage. Animal waste from large feed lots and vegetable waste from canneries provide an enormous amount of organic material. This means that a virtually unlimited supply of nutrients is available.

The technology to convert animal wastes into feed with a series of ponds exists but has not been implemented on a broad scale. Instead, in the feed lots, the manure is scraped into large piles and left to decompose, which means nutrients are lost. If the manure were put into large algae ponds, its nutrients could be recovered in the form of algae, which could then be put to use in a variety of ways.

Since algae is 50-60% protein, it could serve as a feed supplement for animals in the feed lot. Additionally, a portion of the algal culture could be diverted from the algal-bacterial pond to a covered "digester pond," where anaerobic bacteria would ferment the algae into methane gas. The gas would then be "scrubbed" with water (to remove corrosive sulfur compounds) and burned to generate electricity to meet the power needs of the feed lot. The liquid that flows out of the digester after methane fermentation has enough nutrients left in it to be used as a liquid fertilizer. Finally, an additional stream of concentrated algae could be separated from the growth unit for recovery of useful chemicals, such as beta-carotene.

In the waste conversion system designed by Dr. Oswald (see diagram) liquid organic wastes enter a settling tank (1) for removal of floatable and settleable solids. The resulting liquid, rich in dissolved organic matter, is pumped into a pond (2) in which algae supply oxygen for bacteria to oxidize the dissolved organics and release nutrients that promote algal growth. A stream of algae-rich liquid flows into a separation tank (3) in which de-watering (concentration) of algae occurs. Excess water is used for irrigation and concentrated algae for chemical production. A more dilute stream of algae flows into a "digester pond" (4) where anaerobic bacteria ferment algae to methane gas. The gas flows through a "scrubber" (5) and is subsequently burned to generate electricity.

The use of Chlorella to convert waste into clean water and feed is perhaps the best demonstration of Chlorella's alchemical powers of creating the "emerald food."

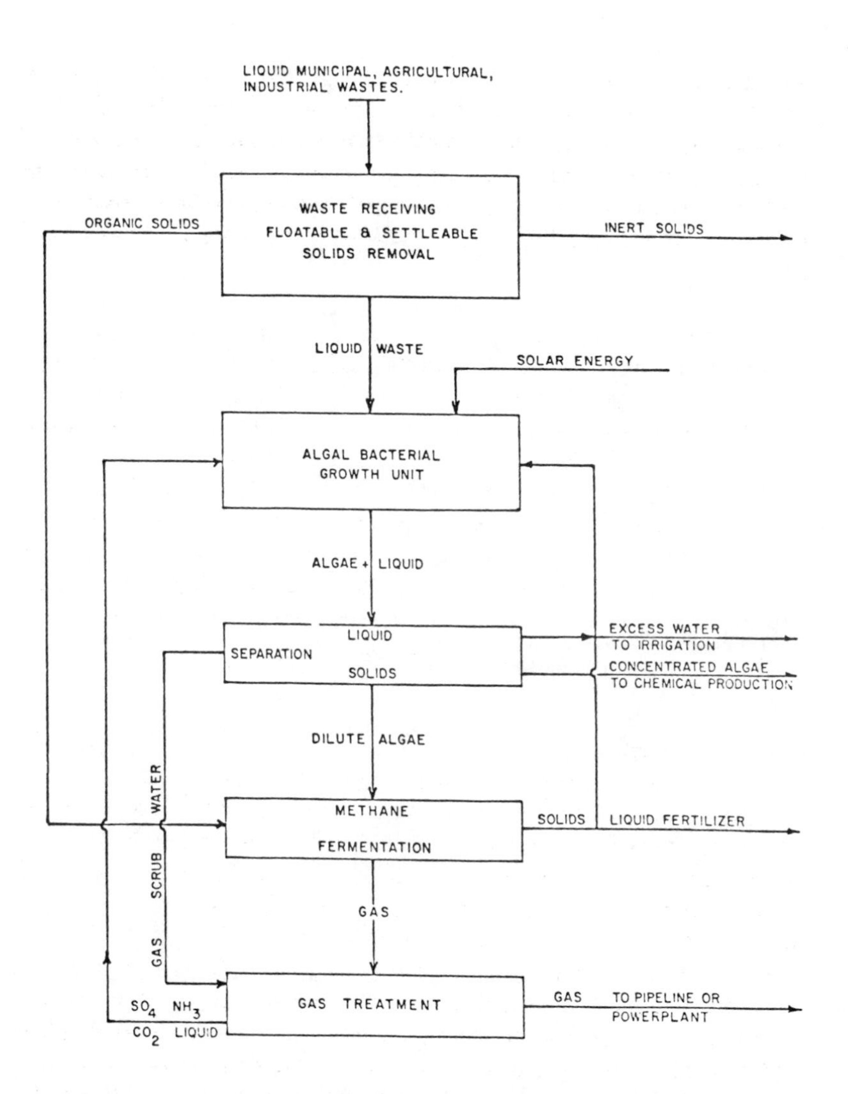

Five step process for converting municipal agricultural and organic industrial wastes to algae, chemicals, fertilizer and natural gas. Courtesy of Dr. William Oswald, University of California at Berkeley.

Microorganisms As Feed and Food Protein Conclusions

Plants with a total capacity for producing 300,000 tons/year of microorganisms for protein from ethanol, methanol, and n-parafins as substrates were operating worldwide in 1976. Another 300,000 ton/year capacity was either in trial production or under construction. Construction of capacity for another 200,000 tons/year was either delayed or stopped (Dimmling and Seipenbusch, 1978). At that point, the future of microorganisms grown on petroleum substrates seemed uncertain.

It would seem, however, that ultimately usage will increase for microbial protein sources. The largest increase will come in livestock feeding with some continued development of these products to supplement the human diet. Although this field of development was given its impetus by the production of various organisms on hydrocarbon media, the growing trend is to use other types for substrates. This is due in no small measure to the problems imposed by the presence of various aromatic polycyclic hydrocarbon residues often found in single cell proteins, even though these have not been shown to be harmful in extensive feeding trials and in fact often occur in higher quantities in naturally occurring products.

Animal feeding studies have generally demonstrated that the various types of single cell proteins may be fed effectively to virtually all types of domestic animals. Physical form and level of inclusion in the diet appear to play almost as great a role in their acceptance by the animal as does their nutritional balance. Additional studies to improve the texture and form of the various products would aid in increased usage.

To date the major research emphasis has been upon the protein or amino acid composition of the various products. The energy contribution has also been recognized and studied extensively although voids in this area are evident. Much more needs to be done to determine the extent and value of other nutritive components of microbial protein, specifically minerals and vitamins. For example, the controversy surrounding the ability of animals to utilize the considerable quantity of phosphorus found in many single cell products should be resolved. More extensive delineation of the nutrient components of the products needs to be made.

For human feeding, the problem of increased nucleic acid ingestion with the accompanying risk of increased blood uric acid levels will inhibit the increased usage of microbial proteins for some time to come. The more conventional single cell products will continue to be modest dietary supplements, but it does not appear likely that these products will make a major contribution to the daily protein needs for some time to come. Techniques have been developed and are avaialble which can markedly reduce the nucleic acid levels; these may be quite useful in the future when greater demand for microorganism products will develop.

Much progress has been made in the development of algae growing systems based on sewage by-products and effluent. Such systems appear capable of reducing some of the problems associated with concentrated animal production as well as contributing a valuable feed ingredient. However, more extensive studies need to be made regarding the accumulation of heavy metals in such systems.

One of the advantages of hydrocarbon substrates for production of single cell proteins is that they can be fairly well defined chemically and production procedures can be reasonably well controlled. As other substrates begin to be used and less elaborate production systems are developed, care must be taken to ensure that production of undesirable or less desirable species does not take place at the expense of organisms whose quality is known. It is well documented that the nutritive quality of the various single cell products is influenced by processing techniques; this may be a limiting factor in making more extensive usage of microbial proteins from less sophisticated systems.

Reprinted by permission from Park W. Waldroup, "Microorganisms As Feed and Food Protein, in Altschul & Wilcke, (eds), *New Protein Foods*, Vol.4, Animal Protein, Part B, Academic Press, New York, 1981, p.244-246

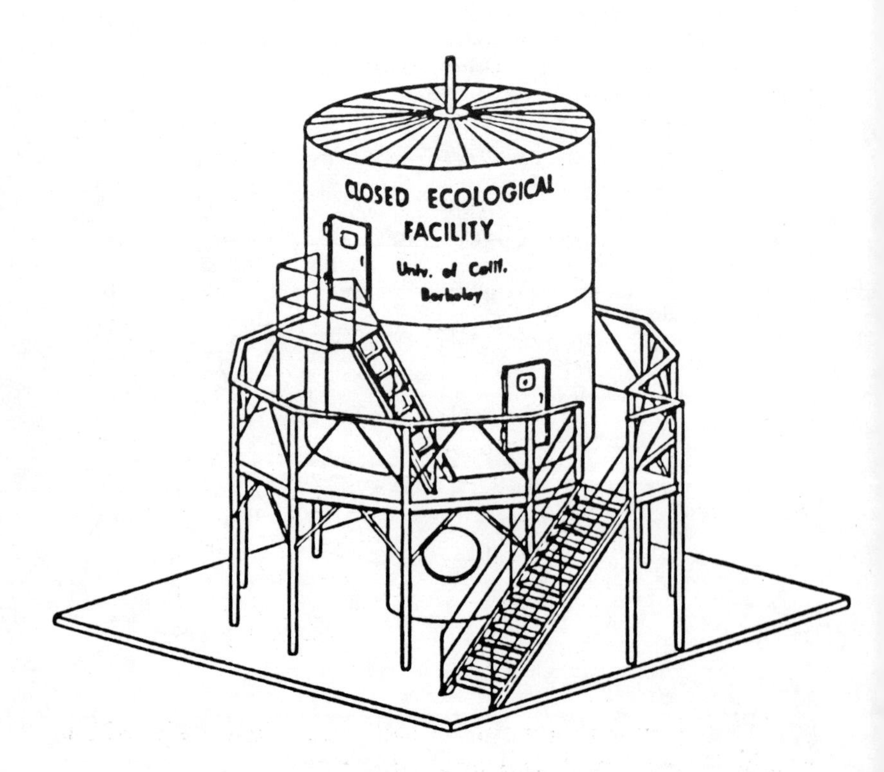
CLOSED ECOLOGICAL
FACILITY
Univ. of Calif.

Chlorella in Space

One of the most intriguing aspects of Chlorella is its potential use for survival in outer space. Among the major problems of space travel are waste removal and obtaining adequate supplies of air, water, and food. There are two basic approaches to these life support problems: carrying along the necessary air, water, and food, and storing the waste for later removal, or creating the necessities for survival while the craft travels through space. The latter approach requires development of a "Controlled Ecological Life Support System" (CELSS), in which waste is recycled and food is grown on board the ship.

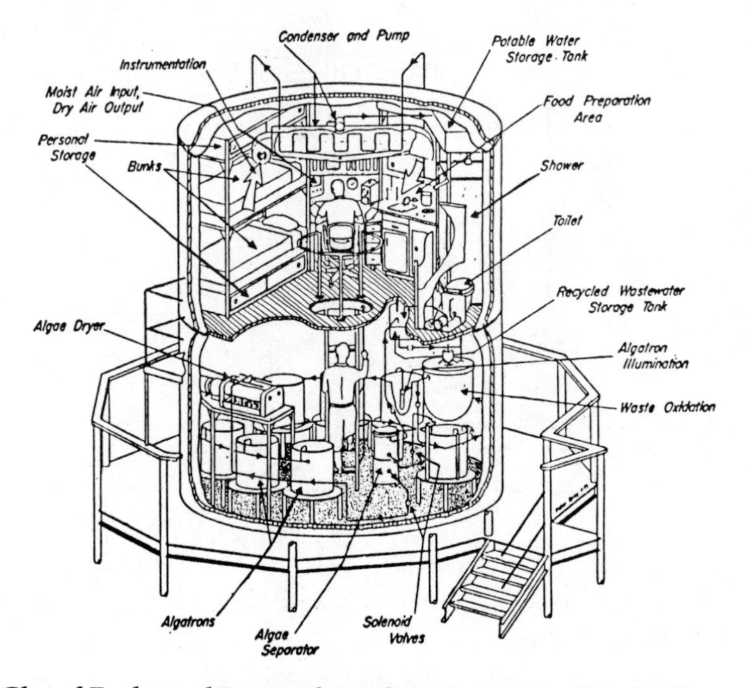

The Closed Ecological Research Facility. Courtesy of Dr. William Oswald, University of California at Berkeley.

In the 1950s, Air Force funded research indicated that algae offer a number of special advantages for use in a CELSS. This research spurred further investigation of the use of Chlorella to create a regenerative life support system in which waste output (carbon dioxide, feces, and urine) is converted into oxygen, clean water, and food — the essentials of human survival.

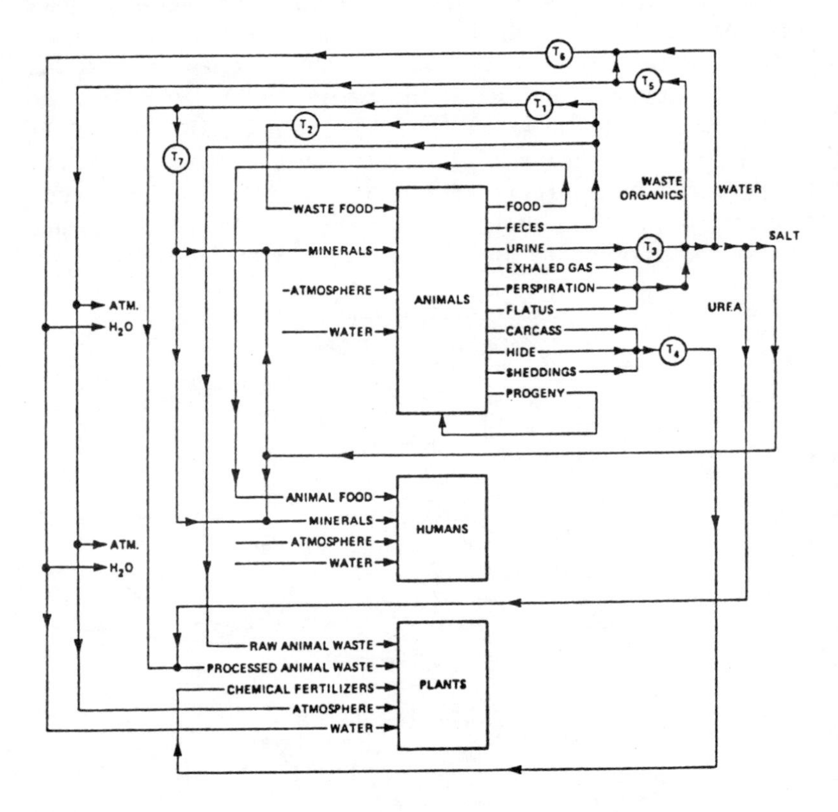

Loop-closing flow sheet for generalized closed life-support system. Courtesy of Dr. William Oswald of University of California at Berkeley.

In the 1960s and early 1970s, Dr. William Oswald demonstrated that 60 liters of Chlorella culture could support the entire metabolism of an adult male. The life support system he pioneered uses algae and bacterial culture grown in an "algatron," a transparent rotating cylinder. Rotation forces the material to spread vertically up the walls of the cylinder, which increases the amount of light energy reaching the culture. Waste material enters the algatron and is broken down by the bacteria and algae, resulting in regeneration of oxygen, water, and food (in the form of new algae).

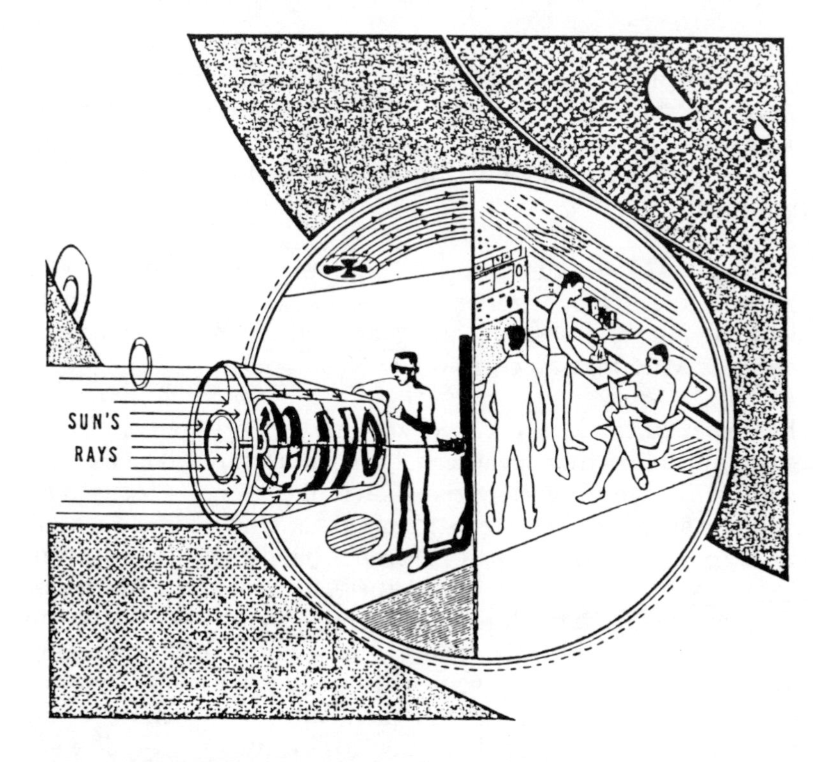

Schematic design of toroidal space station showing algatron in illumination chamber. Courtesy of Dr. William Oswald of University of California at Berkeley.

The possibility of growing algae for food during long space flights encouraged research around the world. After reviewing four major Japanese nutrition publications, three Russian microbiologists, Boykov, Klyushkina, and Kondratyev, concluded that single cell algae can supplement to human nutrition in quantities of 30 to 40 grams per day. Furthermore, findings of Dr. Powell's research team indicated that man can survive on a diet containing up to 100 grams of dry algae for substantial periods of time. All these results pointed to algae as an ideal food for outer space travel.

During the race for the moon in the 1960s, when NASA shifted its emphasis from biological approaches to physical/chemical systems to provide short-term life support for orbital flights and moon shots. The physio-chemical systems offered a good solution because their technology was more advanced and less costly to implement.

In the late 1970s, the prospect of long-term space missions renewed interest in regenerative systems. Scientists began to focus on a hybrid system, one that would combine biological processes with physical/chemical devices.

The Soviet research, as reviewed by Sheryl Bergstrom of NASA, leaves no doubt that algae can be used as the vital link in a regenerative life support system. Current research is focused on designing such hybrid systems to meet changing human needs. A small group asleep in a ship, for example, has different needs from a large crew working outside on the same ship. Mel Avener of NASA is developing methods of using photosynthesis to revitalize the air in the spacecraft taking into account such changing requirements. Developing such a gas exchange process requires ingenuity. When oxygen is too plentiful, toxicity may result, whereas too much carbon dioxide acidosis causes psychic distress and ultimate suffocation. Yet, if the carbon dioxide level is too low, algae and other plants cannot survive.

Once this basic life support problem is solved, NASA is likely to resume work on questions of food production. Dr. Oswald's discoveries in the conversion of waste to clean water and food will undoubtedly play a key role.

The Japanese Chlorella industry has developed the technological ability to create food products of high nutritive value that are both tasty and esthetically pleasing. But one of the stumbling blocks space scientists must overcome, while in flight, is unlocking Chlorella's rich store of nutrients from within its rigid cell wall. Although there have been many advances in this technology, replicating it on a spaceship may present special problems. For this reason, attention has turned recently to another alga, Dunaliella, which has no cell wall. Another problem yet to be solved is the adequate purification of the alga to make it safe for human consumption. Pasteurization would probably be required to destroy bacterial contaminants.

Recreating Spaceship Earth

In space travel, Chlorella and other algae seem to hold the secret of recreating Spaceship Earth. To allow exploration of the outer reaches of the universe, a way must be found to duplicate terrestrial biological processes that will support human life indefinitely. Such a system will replicate the earth's own ecology, involving interaction among all the ship's organisms: the humans, the plants, and the bacteria.

The Idealized Life Support System

Drs. Acker and Stern employed Dr. Oswald's discoveries to conceptualize an idealized algae life support system in which about four pounds of waste yield one and one-half pounds of food. In this system, liquid and solid bodily wastes are fed into a treatment tank where they are decomposed by bacteria. Nutrients released from the waste are fed into algae culture tanks to stimulate algal growth. Gases produced by the bacteria include methane, which can be burned for cooking food on board the craft, and carbon dioxide. This carbon dioxide, as well as that exhaled by the crew, is percolated through the algae tank. The algae use this carbon dioxide in photosynthesis, during which new cells and oxygen are produced. The oxygen is then carried to a dehumidifier and is circulated for the crew to breathe, while the algae are harvested, dried, and used for food. These carbon dioxide/oxygen and waste/food conversions form the basis of a CELSS based on algae.

Eliminating World Hunger

Chlorella and other microalgae can help solve the problem of malnutrition in the world. One-third of the world's population lacks adequate food, and five hundred million people suffer from acute malnutrition. The prodigal productivity of Chlorella and other microalgae may very well help solve this serious problem.

Many micro-algae thrive on arid, unproductive land, under conditions where it is often impossible to grow conventional crops. The searing sun of the desert, and the brackish water (which is often the only water available in such regions), constitute an excellent environment for algo-culture. Today one of the most successful algo-culture systems in the world harvests Dunaliella algae from the concentrated brine of the Dead Sea in Israel. Dunaliella has no cell wall, yet survives in highly saline environments by producing large amounts of glycerol. Glycerol may be used as a sweetener in foods, as a lubricant, or as a building block for the production of more complex chemicals and fuels. When artificially stressed, Dunaliella produces high concentrations of beta-carotene, the precursor for Vitamin A. Recent studies have shown that Vitamin A plays a role in cancer prevention. Unfortunately, large doses of Vitamin A can be toxic to the liver. The advantage of using beta-carotene is that it is not toxic and is converted to Vitamin A by the body only as needed.

In the United States, the first commercial algae plants are now being developed in the inhospitable desert regions of southern California. Tropical countries such as India are well suited for algae cultivation throughout the year. In view of the widespread nutritional deficiency in Peru, and the geographic and climatic conditions that impede production of traditional foodstuffs there, mass culture of microalgae may help ease that country's food shortages. The results of a study by Drs. Castillo, Merino, and Heussler clearly show the ecological appropriateness of algoculture to Peru's arid regions. The prevailing climatic conditions insure high and stable yields, estimated at 80 tons of biomass per hectare yearly. In addition, algo-cultures open the possibility of producing high-quality protein with an efficiency of water utilization far superior to that of traditional crops cultivated on desert terrain. Cultivation techniques can be adapted to local conditions, since the evaporation rate and salinity of the water vary from place to place.

The Poor World Development Group*, a nonprofit organization for improving health, education and commerce in the rural communities of developing countries, has conducted some interesting experiments using Spirulina cultivation in small rural villages as part of a holistic solution to the problems of sanitation, fuel, fertilizer, and food.

---

*The Poor World Development Group depends on the generosity of individuals to carry out its projects. We urge you to make your contribution to the most economical and most efficient form of diplomacy, good will: the kind that goes directly to the villages. Send your tax-deductible donations to the Poor World Development Group, Country Rt. 5, New Lebanon Center, Columbia County, NY 12126, (518) 794-8913.

Malnutrition among villagers results from insufficient protein intake and parasite infestation. One of the main causes of intestinal parasites and other diseases in warm climates is the lack of hygienic facilities for waste disposal. The parasites drain a huge amount of energy from the rural population and consume up to 30% of the food ingested by humans. Many designs for simple latrines are tested in the Poor World Development Group's pilot projects. The parasites are eliminated by the proper handling of sewage. Among the most promising, safe methods of handling sewage are biogas digesters and algae ponds.

The waste digesters, constructed of thin-walled ferrous cement, use thermal destruction as the first step in eliminating disease-producing microorganisms. With solar heating, a digester operates at about 55 degrees Centigrade. At this temperature the thermophilic ("heat-loving") bacteria replace the disease-causing microorganisms. This is the first of a series of barriers to the spread of disease between the sanitation system and the edible algae. Dr. William Oswald whose pioneering research serves as the foundation of most of these systems has serious reservations about the advisability of consuming algae grown of animal waste.

The waste from plants, latrines, and livestock is pumped into "the waste digester" containing bacteria, which produce liquid effluent, biogas, and solid wastes. The solids are made into compost. The biogas is separated into methane for burning and carbon dioxide for the algae pond. The liquid effluent is filtered and used as fertilizer for the fields or for the algae pond. The algae basin converts waste to algal mass. The algae are harvested and dried with solar energy and produce a nutritious powder rich in proteins and vitamins.

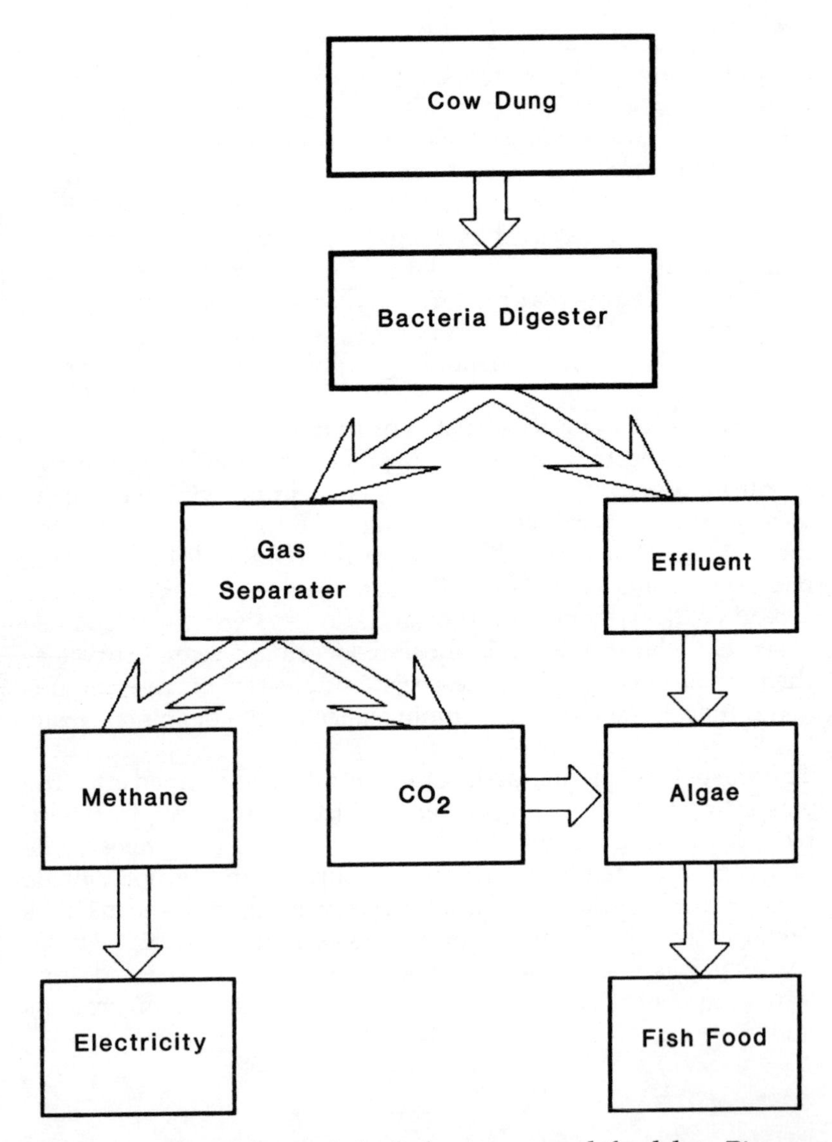

Using Dr. Oswald's model of the integrated feed lot, Pinnan Soong in Taiwan developed a model for an integrated farm. Here, cow dung was converted to methane gas and algae. The methane was burned to produce electricity. The algae was fed to fish. Villagers then ate the fish.

In India Ripley Fox helps villages become self-sufficient by using self-contained production facilities such as this raceway pond. Photo courtesy of Earthrise™ Spirulina.

We have yet to see algo-culture systems producing tons of high-grade protein in the poor Third World countries where it is most needed. Perhaps the most important reason is that technology for harvesting these foods is still in the experimental stage. This means that constant quality control monitoring, and the high cost of research and development, add tremendously to production costs. At present it is still more expensive to produce protein by means of algo-culture than through conventional systems of agriculture.

Fortunately, protein is only one of the significant components of Chlorella. Interest in this new food source is growing on a world-wide basis. Sales of Chlorella and other microalgae in the health food markets of the wealthier nations are now providing financing and economic incentives for further research. In the future this will lead to lower production costs and ultimately to operating systems capable of turning out tons of this proteinand vitamin-rich food where it is most needed —not just as a health-enhancing supplement, but for survival.

The Amazing Alchemist Bibliography

Abbott, W.E., Oxygen Production in Water by Photosynthesis, *Sewage Works Journal*, 1948, v.20, p.538

Acker, J.A. and J.A. Stern, WADD Technical Report 60-574, Aerospace Medical Research Laboratory, Wright-Patterson AFB, Ohio, 1960, p.179-196

Altschul, Aaron M. and Harold L. Wilcke (eds), *New Protein Foods: Animal Protein Supplies*, Part B, Academic Press, New York, 1981

The Art of Growing Premium Quality Spirulina, *Earthrise Newsletter*, 1983, Apr, v.3, n.1

Averner, Maurice M., An Approach to the Mathematical Modelling of a Controlled Ecological Life Support System, NASA Contractor Report 166331, 1981

Barak, Amnon, Research and Development in Applied Algology and the Protein Shortage Problem, *Algae Biomass*, Elsevier/North-Holland Biomedical Press, 1980, p.21-23

Bardach, J.E., *Aquaculture Science*, 1968, v.161, p.1098-1112

Becker, E.W., The Legislative Background For Utilization of Microalgae and Other Types of Single-Cell Protein in Soder and Binsack (eds), *Microalgae For Food And Feed*, Ergebn. Limnol, 1978, v.11, p.56-64

Becker, E.W. and L.V. Venkataraman, Production and Processing of Algae in Pilot Plant Scale: Experiences of the Indo-German Project in *Algae Biomass*, Elsevier/North-Holland Biomedical Press, 1980, p.35-50

Becker, E.W., L.V. Venkataraman and P.M. Khanum, Digestibility Coefficient and Biological Value of the Proteins in The Algae in *Scenedesmus Acutus*, Processed by Different Methods, *Nutr. Rep. International*, 1976, v.14, n.4

Becker, M.J. and A.M. Shefner, Research on the Chemical Composition and Digestibility of Algal Cell Walls, AMRL-TDR-63-115, Wright-Patterson AFB, Ohio, 1963

Bellows, Bob, Spirulina: Nature's Magic Algae, *Muscle & Fitness*, p.83-84

Ben-Amotz, Ami and Mordhay Auron, Glycerol, B-Carotene and Dry Algal Meal Production by Commercial Cultivation of Duniella in Shelef and Soeder (eds), *Algae Biomass*, Elsevier/North Holland Biomedical Press, 1980, p.603-610

Bender, A.E. (ed), *Evaluation Of Novel Protein Products*, Perganon Press

Benemann, J.R., Biofuels: A Survey, *Electric Power Research*, June 1978

Benemann, J.R., Hollaender (ed), *The Biosaline Concept*, Plenum Press, New York, 1979

Benemann, J.R., Energy Journal, 1980, v.1, p.107

Benemann, J.R., J.C. Weissman, and W.O. Oswald, in Rose (ed), *Microbial Biomass*, Academic Press, New York, 1979

Benemann, John and Joseph Weissman, Chemicals from Microalgae in Wise (ed), *Bioconversion Systems*, CRC Series in Bioenergy Systems, CRC Press, Boca Raton, FL, 1984, p.60-70

Bergstrom, S.L., *Regenerative Life Support Systems: A Survey of Soviet Accomplishments*, NASA, Jet Propulsion Laboratory, California Institute of Technology, Pasadena

Bitton, G., J.R. Fox and H.G. Stuckland, Removal of Algae from Florida Lakes by Magnetic Filtration, *Applied Microbiology*, 1975, v.30, n.6, p.905-908

Bongers, L., *Aerospace Med.*, 1964, v.35, p.139

Bongers, L., and K. Kock, *Develop.Industrial Microbiology*, 1964, p.183

Bovee, H.H., A.J. Pilgrim, L.S. Sun, J.E. Schubert, T.L. Eng, and B.J. Benishek, Biologistics for Space Systems Symposium AMRL-TDR-62-116, Aerospace Medical Research Laboratory, Wright-Patterson AFB, Ohio, 1962, p.8-19

Boykov, N.N., N.S. Klyushkina, Yu. I. Kondrat'yev, Vopr. Pitaniya, 1963, v.21, n.5

Boykov, N.N., V.P. Bychkov, Yu. I. Kondrat'yev, and A.S. Ushakov, Vopr. Pitaniya, 1962, v.21, n.5, p.76

Boykov, N.N., N.S. Klyushkina, and Yu. I. Kondrat'yev, Vopr. Pitaniya, 1963, v.22, n.6, p.3

Bremer, H.J., Hazards and Problems in the Utilization of Microalgae For Human Nutrition in Soeder and Binsack (eds), *Microalgae For Food and Feed*, Ergebn. Limnol., 1978, v.11, p.218-222

Brockman, M.D., A.S. Henick, G.W. Kurtz, and R.G. Tischer, Bibliography, *Food Technology*, 1958, v.12, p.449-453

Buderer, M.D., Russian Biospex: Biological Space Experiemnts, A Space Life Sciences Bibliography, Prepared under Contract NAS 9-15850, JSC -17072, 1981

Buri, P., and W. Yongmanitchai, A New CO2 Supply System (Heussler's System) For Shallow Water Outdoor Mass Algal Culture, Algae Project, Kasetsart University, Bangkok, 1977, p.154-161

Burlew, J.D. (ed), *Algal Culture: From Laboratory to Pilot Plant*, Carnegie Institute, Washington, D.C., n.600, 1955

Byrne, A.F., and R.B. Koch, *Science*, 1962, v.135, p.215

Caldwell, D.H., Sewage Oxidation Ponds: Performance Operations and Design, *Sewage Works Journal*, 1946, v.18, p.433

Calloway, Doris Howes, The Place of SCP in Man's Diet, in Davis (ed), *Single-Cell Protein*, Academic Press, New York, 1974, p.129-146

Calloway, D.H., and S. Margen, NASA Grant NGR-05-003-089, NASA Headqrtrs., Washington, D.C., 1965

Carden, J.L. and R. Browner, Preparation and Analysis of Standardized Waste Samples for Controlled Ecological Life Support Systems (CELSS), NASA Contractor Report 166392, 1982

Casey, R.P., and J.A. Lubitz, Food For Space, *Food Technology*, 1963, v.17, p.1386

Castillo, J.S., M.F. Merino and P. Heussler, Production and Ecological Implications of Algae Mass Culture Under Peruvian Conditions in *Algae Biomass*, Elsevier/North Holland Biomedical Press, 1980, p.123-134

Chapman, D.D., R. Meyers, and C.M. Proctor, *Devel. Industrial Microbiology*, 1962, v.3, p.343

Cieeri, Orio, Spirulina: The Edible Microorganism, *Microbial Reviews*, American Society for Microbiology, 1983, v.47, n.4, p.551-578

Clement, Genevieve, A New Type of Food Algae in Mateles and Tannenbaum (eds), *Single-Cell Protein*, American Society for Microbiology, 1983, v.47, n.4, p.551-578

Cook, B.B., E.W. Law, and B.M. Bailey, *Journal of Nutrition*, 1963, v.81, p.23

Cronshaw, J., A. Myers, and R.D. Preston, *Biochem. Biophys. Acta*, 1958, v.27, p.89

Dabah, R., Protein From Microorganism, *Food Technology*, 1970, v.24

Dam, R., L. Lee, P.C. Fry, and H. Fox, Utilization of Algae As Protein Source For Humans, *Journal of Nutrition*, 1965, v.86, p.376-382

Davis, P. (ed), *Single-Cell Protein*, Proceedings of the International Symposium, Rome, Italy, 1973, Nov, Academic Press, New York, 1974

Dixon, Bernard, *Magnificent Microbes*, Atheneum, New York, 1976

Dodd, J.C. and J.L. Anderson, An Integrated High Rate Bond Algae Harvesting System, *Prog. Water Technology*, 1977, v.9, p.713-726

Dubinsky, Z., and Aaronson, S., Review of the Potential Uses of Microalgae in San Pietro (ed), *Biosaline Research: A Look At The Future*, Plenum Press, NY, 1982, p. 181-206

Dugan, G.L., C.G. Golueke, and W.J. Oswald, Recycling System for Poultry Wastes, *Journal Water Pollution Control Federation*, 1972, v.44, n.3, p.432-440

Durand-Chastel, H., Production and Use of Spirulina in Mexico, in Shelef and Soeder (eds), *Algae Biomass*, Elsevier/North Holland Biomedical Press, 1980, p.51-63

Eckenfelder, W.W. and Brother Joseph McCabe, *Biological Waste Treatment*, Pergamon Press, London, 1961

Endo, H., H. Sansawa and K. Nakarjuna, Studies on *Chlorella Regularis*, Heterotrophic Fast Growing Strain II, *Plant and Cell Physics*, 1977, v.18, p.199-208

Enebo, L., *Evaluation of Novel Protein Sources*, Pergamon Press, New York, 1968, p.93-103

Fofanov, V.I., M.I. Kozar, and N.N. Dobronravova, Some Indices of the Human Natural Resistance to the Dietary Replacement of Animal Protein by Chlorella Proteins, *Kosmich. Biol. I. Med.*, 1967, v.1, n.3, p.121-127

Folkman, Yair and A.M. Wachs, Removal of Algae From Stabilization Pond Effluents by Lime Treatment, *Water Research*, 1973, v.7, n.3, p.419-435

Fong, F., and E.A. Funkhouser, Air Pollutant Production by Algal Cell Cultures, NASA Contractor Report 166384, 1982

George, Uwe, *In The Deserts Of This Earth*, Harcourt Brace Jovanovich, 1977

Gitel'zon, I.I. (ed), Problems of Creating Biotechnical Systems of Human Life Support, (Translated from Russian), NASA Technical Translation F-17533, 1975

Gitel'zon, I.I., B.G. Kovrox, G.M. Lisovskiy, Yu. N. Okladnikov, M.S. Rerberg, F. Ya. Sidko and I.A. Terskov, Experimental Ecological Systems Including Man in *Problems Of Space Biology*, (Translated from Russian), NASA Technical Translation F-16993, 1975, v.28

Gloyna, E.F. and W.W. Eckenfelder, Jr., *Advances in Water Quality Improvement*, University of Texas Press, 1968

Goldman, J.C., W.J. Oswald and D. Jenkins, The Kinetics of Inorganic Carbon Limited Algal Growth, *Journal Water Pollution Control Federation*, 1974, v.46, n.3, p.554-574

Golueke, C.G. and W.J. Oswald, Harvesting and Processing Sewage-Grown Planktonic Algae, *Journal of Water Pollution Control Federation*, 1965, v.37, n.4, p.471-498

Golueke, C.G., W.J. Oswald and H.B. Gotaas, Anaerobic Digestion of Algae, *Applied Microbiology*, 1957, v.5

Golueke, C.G. and W.J. Oswald, Surface Properties and Ion Exchange in Algae Removal, *Journal of the Water Pollution Control Federation*, 1970, Aug, v.42, n.8

Golueke, C.G. and W.J. Oswald, Large-Scale Production of Algae in Mateles and Tannenbaum (eds), *Single-Cell Protein*, M.I.T. Press, Cambridge, MA, 1968

Golueke, C.G., W.J. Oswald, and P.H. McGauhey, Biological Control of Enclosed Environments, *Sewage and Industrial Wastes*, 1959, v.31, n.10, p.1125-1142

Golueke, C.G., H.B. Gotaas, and W.J. Oswald, Recovery of Algae From Waste Stabilization Ponds, *Sanitary Engineering Research Laboratory Series*, 1957-58, v.44, n.7,8

Gotaas, H.B., H.F. Ludwig and V. Lynch, Growth Characteristics of Chlorella Pyrenoidosa Cultured in Sewage, *Sewage and Industrial Wastes*, 1953, Jan, v.25, n.1

Grau, C.R. and N.W. Klein, *Poultry Science*, 1957, v.36, p.1046

Gray, W.D. and M.A. El Seoud, *Develop. Industrial Microbiology*, 1965, v.7, p.221

Grobbelaar, J.V., Observations on The Mass Culture of Algae As A Potential Source of Food, *South Africa Journal of Science*, 1979, v.75, p.133-136

Hayami, H., Y. Matsuno, and K. Shino, Studies on the Utilization of Chlorella as a Source of Food (Part 8), *Annual Report National Institute Nutrition*, Japan, 1960, p.58

Hayami, H. and K. Shino, Nutritional Studies on Decolorized Chlorella, (Part 2), *Annual Report National Institution Nutrition*, Japan, 1958, p.59

Hayami, H., K. Shino, K. Morimoto, T. Okano, and S. Yamamoto, Studies on the Utilization of Chlorella as a Source of Food, (Part 9), *Annual Report National Institute Nutrition*, Japan, 1960, p.60

Heussler, P., Improvements in Pond Construction and CO2 Supply For the Mass Production of Microalgae in Soeder and Binsack (eds), *Microalgae For Food And Feed*, Ergebn. Limnol., 1978, v.11, p.254-258

Hills, Dr. Chrisopher and N. Hiroshi, *Food From Sunlight*, University of the Trees Press, 1978

Hintz, H.F., H. Heitman Jr., W.C. Weir, D.T. Torell, and J.H. Meyer, Nutritive Value of Algae Grown on Sewage, *Journal of Animal Science*, 1966, v.25, n.3, p.675-681

Hollaender, A. (ed), *Trends In The Biology of Fermentation For Fuels And Chemicals*, Plenum Press, New York, 1981

Hooper, M.W., Disposal of Wastes from Vessels, *Oceanology International*, 1971, Sep, p.37-38

Huffaker, R.C., D.W. Rains, and C.O. Qualset, Utilization of Urea, Ammonia, Nitrite, and Nitrate by Crop Plants in a Controlled Ecological Life Support System(CELSS), NASA Contractor Report 166417, 1982

Ives, K.J., The Significance of Surface Electric Charge on Algae in Water Purification, *Journal of Biochemistry and Microbiol. Technology and Engineering*, 1959, v.1, p.37

Jagow, R.B., and R.S. Thomas (eds), Study of Life-Support Systems for Space Missions Exceeding One Year in Duration, General Dynamics, NASA Contract NAS 2-3011, 1966

Jaleel, S.A. and C.J. Soeder, Current Trends in Micro Algal Culture as Protein Source in West Germany; Food and Feed Applications, *Indian Food Packer*, 1973, v.27, n.45

Jensen, Bernard and D.C. Goldman in Leslie (ed), *Health Magic Through Chlorophyll*

Johnston, H.W., The Biological and Economic Importance of Algae, 3. *Edible Algae of Fresh and Brackish Waters*, Tuatara, 1970, v.18, p.19-35

Jorgensen, J. and J. Convit, Cultivation of Complexes of Algae With Other Freshwater Microorganisms in Burlew (ed), *Algal Culture: From Laboratory To Pilot Plant*, Carnegie Institute, Washington, D.C., n.600, 1955, p.190-196

Kandatsu, M. and T. Yasui, *Japan Soc. Food Nutrition*, 1964, v.16, p.516

Kawada, S., Y. Matsuno, K. Watanabe, and T. Ohta, Nutr. Studies on Chlorella, *Annual Report National Institute Nutrition*, Japan, 1960, v.21, p.54-55

Kawaguchi, Kotaro, Microalgae Production Systems in Asia, *Algae Biomass*, Elsevier/North-Holland Biomedical Press, 1980, p.25-33

Kennedy, R., The Future Foods Revolution, *Total Health*, 1981, Nov, p.13-14

Ketchum, B.H. and C. Redfield, *Journal Cell. Comp. Physiology*, 1949, v.33, p.281

Kirenskiy, L.V., I.A. Terskov, I.I. Gitel'zon, G.M. Lisovskiy, B.G. Kovrov, Y.N. Okladnikov, Experimental Biological Life Support System, Part II, Gas Exchange Between Man and Microalgae Culture in a 30-Day Experiment, Brown and Favorite (eds), *Life Sciences and Space Research*, North-Holland Publishing Company, Amsterdam, 1967, p.37-40

Konecci, E.B., *Space Science Reviews*, 1966, v.6, p.3

Krauss, R.W., Conference on Nutrition in Space and Related Waste Problems, NASA, SP-70, Washington, D.C., 1964, p.289-297

Krauss, R.W., *American Journal of Botany*, 1962, v.49, p.425

Lachance, P.A., *Nutrition News*, 1966, v.29, p.13

Lachance, P.A., Conference on Nutrition in Space and Related Waste Problems, NASA, SP-70, Washington, D.C., 1964, p.71-78

Lachance, P.A., Single-Cell Protein in Space Systems in Mateles and Tannenbaum (eds) *Single-Cell Protein*, The M.I.T. Press, Cambridge, MA, 1968, p.122-152

Lachance, P.A. and C.A. Berry, *Nutrition Today*, 1967, v.2, p.2

Lachance, P.A. and J.E. Vanderveen, *Food Technology*, 1963, v.17, p.59

Laskin, A.I. and H.A. Lechevalier (eds), *Handbook Of Microbiology*, CRC Press, Boca Raton, FL, 1978, p.348

Lee, S.K., H.M. Fox, C. Kies, and R. Dam, The Supplementary Value of Algae in Human Diets, *Journal of Nutrition*, 1967, v.92, p.281-285

Leveille, G.A., H.E. Sauberlich, and J.A. Edelbrock, U.S. Army Medical Research & Nutrition Laboratory Report, 1961, p.259

Leveille, G.A., H.E. Sauberlich, and J.W. Shockley, *Journal of Nutrition*, 1962, v.76, p.423

Lind, S.C. and D.C. Bardwell, *Journal of the American Chemical Society*, 1926, v.48, p.2335

Lubitz, J.A., Biologistics for Space Systems, Symposium, AMRL-TDR-62-116, Aerospace Medical Research Laboratory, Wright-Patterson AFB, Ohio, 1962, Oct, p.331-356

Ludwig, H.F. and W.J. Oswald, Role of Algae in Sewage Oxidation Ponds, *Scientific Monthly*, 1952, v.74, p.3-6

Mason, R.M. and J.L. Carden (eds), Controlled Ecological Life Support System: Research and Development Guidelines, NASA Conference Publication 2232, 1982

Mateles, R.I. and S.R. Tannenbaum (eds), *Single-Cell Protein*, The M.I.T. Press, Massachusetts Institute of Technology, 1968

Mateles, R.I., J.N. Baruah, and S.R. Tannenbaum, *Science*, 1967, v.157, p.1322

McDowell, M.E. and G.A. Leveille, Feeding Experiments With Algae, *Fed.Proceedings*, 1963, v.22, p.1431-1438

McGarry, M.G., Algal Flocculation with Aluminum Sulfate and Polyelectrolytes, *Journal Water Pollution Control Federation*, R191-R201, 1970, v.42, n.5, p.191-201

McGauhey, P.H., *Engineering Management Of Water Quality*, McGraw-Hill, 1968

McPherson, A.T., *Journal of Animal Science*, 1966, v.25, p.575

Merrell, J.C., J.R, and A. Katko, Reclaimed Wastewater for Santee Recreational Lakes, *This Journal*, 1966, Aug, v.38, n.8, p.1310

Mikeladze, G.G., R.G. Meskhi, Yu. F. Zav';yalov, N.I. Kobaidze and N. Ye. Lortikipanidze, Use of *Spirulina* Biomass as Food Together With Higher Plants, Kordyum (ed) in *Role of Lower Organisms In Recycling Of Substances In Closed Ecological Systems*, NASA Technical Memorandum TM-76484, 1979, p.43-47

Miller, R.L. and C.H. Ward, Algal Bioregenerative Systems in Kammermeyer (ed), *Atmosphere In Space Cabin*, Appleton-Century-Croft Publishing Company, 1966

Milov, M.A. and K.A. Balakireva, Selection of Higher Plant Cultures for Biological Life Support System, Gitel'zon (ed) in *Problems Of Creating Biotechnical Systems Of Human Life Support*, (Translated from Russian), NASA Technical Translation F-17533, 1975, p.13-20

Miyoshi, T., Studies on the Digestion and Absorption of Algae *Chlorella, Scenedemus, Ikikoku Acta Med.*, 1959, v.15, p.1237

Modell, M., Sustaining Life in a Space Colony, *Technology Review*, 1977, Jul-Aug

Mohn, F.H., Improved Technology for the Harvesting and Processing of Microalgae and Their Impact on Production Costs, in Soeder and Binsack (eds), *Microalgae For Food and Feed*, Ergebn. Limnol., 1978, v.11, p.228-253

Moore III, B., R.D. MacElroy (eds), *Controlled Ecological Life Support System: Biological Problems*, NASA Conference Publication 2233, 1982

Morimura, Yuji and T. Nobuko, Preliminary Experiments in the Use of Chlorella as a Human Food, *Food Technology*, 1954, v.3, n.4, p.179-182

Myers, J., Conference on Nutrition in Space and Related-Waste Problems, NASA, 1964, SP-70, p.283-287

Nilovskaya, N.T. and M.M. Bokovaya, Study of Regeneration of Air and Water by Higher Plants in a Closed Space, *Problems Of Space Biology*, 1967, v.7, p.496-506

Northcote, D.H., K.J. Goulding, and R.W. Horne, *Biochemistry Journal*, 1958, v.70, p.391

Oro, J., *Nature*, 1963, v.197, p.862

Oswald, W.J., *Advances in Biological Waste Treatment*, Perganon Press, New York, 1903, p.357

Oswald, W.J., Advances in Environmental Control Studies Wtih A Closed Ecological System, *American Biology Teacher*, 1963, Oct, v.25, n.6

Oswald, W.J., Algal Production Problems, Achievements, and Potential, *Algae Biomass*, Elsevier/North Holland Biomedical Press, 1980

Oswald, W.J., The Coming Industry of Controlled Photosynthesis, *American Journal of Public Health*, 1962, v.52, n.2

Oswald, W.J., Current Status of Microalgae From Wastes, *Chemical Engineering Progress Symposium Series* , 1969, v.65, n.93

Oswald, W.J., Discussion of an Integrated High-Rate Pond Algae-Harvesting System, *Progress in Water Technology*, Pergamon Press, 1977, v.9, p.713-726

Oswald, W.J., The Engineering Aspects of Microalgae in Laskin and Lechevalier (eds), *Handbook Of Microbiology*, C.R.C. Press, West Palm Beach, FL, 1977

Oswald, W.J., Gas Production From Microalgae *Clean Fuels From Biomass And Wastes*, Institute of Gas Technology, Chicago, Il, 1976, Jan, p.311-324

Oswald, W.J., High-Rate Pond in Waste Disposal, *Developments In Industrial Microbiology*, 1963, v.4

Oswald, W.J., Light Conversion Efficiency of Algae Grown in Sewage, *Journal of the Sanitary Engineering Division*, 1960, July

Oswald, W.J., *Microbiological Waste Nutrient Recycle Systems Bulletin*, New Mexico Academy of Science, 1972, Dec, v.13, n.2, p.30-32

Oswald, W.J., A Miniature System for Ecological Research (Closed Environmental Facility), Proceedings of 17th Annual Meeting and Equipment Exposition, Institute Environmental Studies, 1971, April

Oswald, W.J., Photosynthetic Single Cell Protein in *Protein Resources And 1978Logy: Status And Research Needs*, NSF-MIT Prot. Resources Study, Avi Publishing, Westport, CT

Oswald, W.J., Pollutant and Waste Removal from Biosaline Environments, *Biosaline Research: A Look To The Future*, Plenum Press, 1981

Oswald, W.J., Production of Chlorella and Its Bearing on Waste Water Treatment, Presented at the Symposium Industrial Waste Disposal, XIX Congress of International Union for Pure and Applied Chem,., London, U.K., 1963, July

Oswald, W.J., *Progress In Water Technology, Water Quality Management And Pollution Control: Complete Waste Treatment In Ponds*, Pergamon Press, London, 1973

Oswald, W.J., Solar Energy Fixation With Algal Bacterial Systems, *Compost Science*, 1974, Jan-Feb, v.20-21

Oswald, W.J., A Syllabus on Waste Pond Fundamentals, Biomed. & Environmental Health Science, School of Public Health, University of California, Berkeley, 1983

Oswald, W.J. and J.R. Benemann, Algae-Bacterial Systems in *Biochemical And Photosynthetic Aspects Of Energy Production*, Academic Press, 1980, p.59-80

Oswald, W.J. and J.R. Benemann, *Biochemical And Photosynthetic Aspects Of Energy Production*, Academic Press, New York, 1978

Oswald, W.J. and J.R. Benemann, Critical Analysis of Bioconversion With Microalgae, *Biol. Solar Energy Conversion*, Academic Press, 1977, p.379-396

Oswald, W.J. and J.R. Benemann, Fertilizer From Algal Biomass, *Proceedings 2nd Symposium on Research to National Needs*, 1976, Nov, p.29-31

Oswald, W.J. and J.R. Benemann, Freshwater Algae Farming, Conference on Capturing the Sun Through Bioconversion, *Metropolitan Studies*, Proceedings, Washington Center, 1976, March, p.247-248

Oswald, W.J., J.R. Benemann, and B.L. Koopman, Production of Biomass From Freshwater Aquatic Systems, *Concepts of Large-Scale Bioconversion Systems Using Microalgae*, Proc. of Fuels from Biomass Symposium, University of Illinois, 1977, Apr, p.59-81

Oswald, W.J., J.R. Benemann, B.L. Koopman, D.C. Baker, and J.C. Weissman, A Systems Analysis of Bioconversion with Microalgae, *Clean Fuels From Biomass and Wastes*, Institute of Gas Technology, Chicago, Il, 1977, Jan, p.103-126

Oswald, W.J., J.R. Benemann, B.L. Koopman, and J.C. Weissman, Biomass Production and Waste Recycling With Blue-Green Algae in Schlegel and Barnea (eds), *Microbial Energy Conversion*, Goltz, Gottingen, Germany, 1976, Oct, p.413-426

Oswald, W.J., J.R. Benemann, J.C. Weissman, Energy Production by Microbial Photosynthesis, *Nature*, 1977, July, n.268, p.19-23

Oswald, W.J., J.R. Benemann, J.C. Weissman, and N.E. Grisanti, Algal Single-Cell Protein, *Economic Microbiology*, Academic Press, London, 1977, v.4

Oswald, W.J., G.L. Dugan, and C.G. Golueke, Recycle System For Poultry Wastes, *Journal of the Water Pollution Control Federation*, 1972, v.44, n.3, p.437

Oswald, W.J. and Don M. Eisenberg, Biomass Generation Systems as an Energy Resource, *Proceedings Bioenergy 80th Conference*, Atlanta Bioenergy Council, Washington, D.C., 1980

Oswald, W.J., D.M. Eisenberg, J.R. Benemann, R.P. Goebel, and T.T. Tiburzi, Methane Fermentation of Microalgae, *Proceedings of the First International Symposium on Anaerobic Digestion*, University Colleges, Cardiff, U.K., 1979

Oswald, W.J. and C.G. Golueke, An Algal Regenerative System for Single-Family Farms and Villages, *Compost Science Journal of Waste Recycling*, 1973

Oswald, W.J. and C.G. Golueke, The Algatron, A Novel Microbial Culture System, *The Sun At Work*, 1966, Jan, v.11, n.1

Oswald, W.J. and C.G. Golueke, Algae Production From Waste, *Proceedings of the 18th Annual California Animal Industrial Conference*, Fresno, Ca, 1965, Oct

Oswald, W.J. and C.G. Golueke, Biological Transformation of Solar Energy, *Advances In Applied Microbiology*, 1960, v.2, p.223-262

Oswald, W.J. and C.G. Golueke, Biological Control of Enclosed Environments, *Sewage and Industrial Wastes*, 1959, Oct, v.31, n.10

Oswald, W.J. and C.G. Golueke, Biological Conversion of Light Energy to the Chemical Energy of Methane, *Applied Microbiology*, 1959, July, v.7, n.4

Oswald, W.J. and C.G. Golueke, Closed Ecological Systems, *Journal of the Sanitary Engineering Division*, 1965, Aug, v.91, n.SA4

Oswald, W.J. and C.G. Golueke, Closing an Ecological System Consisting of a Mammal, Algae, and Non-Photosynthetic Microorganisms, *American Biology Teacher*, 1963, Nov, v.25, n.7

Oswald, W.J. and C.G. Golueke, Environmental Control Studies With A Closed Ecological System, *Proceedings of the Institute of Environmental Sciences*, 8th Annual Meeting, 1962, p.183-191

Oswald, W.J. and C.G. Golueke, Fundamental Factors in Waste Utilization in Isolated Systems, *Developments in Industrial Microbiology*, 1964, v.5, p.196-206

Oswald, W.J. and C.G. Golueke, Harvesting and Processing of Waste-Grown Micro-Algae in Jackson (ed) *Algae, Man and the Environment*, 1968

Oswald, W.J. and C.G. Golueke, Large-Scale Production of Algae in Mateles and Tannenbaum (eds) *Single-Cell Protein*, M.I.T. Press, Cambridge, MA, 1968

Oswald, W.J. and C.G. Golueke, Man in Space: He Takes Along His Wastes Problem, *Wastes Engineering*, 1961, Sep, v.32, n.9

Oswald, W.J. and C.G. Golueke, Power From Solar Energy Via Algae-Produced Methane, *Solar Energy*, 1963, July, v.7, n.3

Oswald, W.J. and C.G. Golueke, Role of Plants in Closed Systems, *Annual Review of Plant Physiology*, 1964, v.15

Oswald, W.J. and C.G. Golueke, Solar Power Via A Botanical Process, *Mechanical Engineering*, 1964, Feb, v.86, n.2

Oswald, W.J. and C.G. Golueke, Surface Properties and Ion Exchange in Algae Removal, *Journal of the Water Pollution Control Federation*, 1970, Aug, v.42, n.8

Oswald, W.J., C.G. Golueke, R.C. Cooper, H.K. Gee, and J.C. Bronson, Water Reclamation, Algal Production and Methane Fermentation in Waste Ponds, *Journal of International Air and Water Pollution*, 1963, Aug, v.7, n.6

Oswald, W.J., C.G. Golueke, and H.K. Gee, Harvesting and Processing Sewage-Grown Planktonic Algae, *Journal of the Water Pollution Control Federation*, 1965, Apr, v.37, n.4

Oswald, W.J., C.G. Golueke, and H.B. Gotaas, Anaerobic Digestion of Algae, *Applied Microbiology*, 1957, Jan, v.5, n.1

Oswald, W.J., C.G. Golueke, D.O. Horning, G. Shelef, and M.W. Lorenzen, Spinning Chemostats for Mass Cultures of Microorganisms, *Proceedings of the American Institute of Chemical Engineers*, Series 86, 1968, v.64

Oswald, W.J., C.G. Golueke, and R.W. Tyler, Integrated Pond Systems for Subdivisions, *Journal of the Water Pollution Control Federation*, 1967, Aug, v.39, n.8

Oswald, W.J. and H.B. Gotaas, Discussion: Photosynthesis in the Algae, *Industrial and Engineering Chemistry*, 1956, Sep, v.48, n.9

Oswld, W.J. and H.B. Gotaas, Utilization of Solar Energy for Waste Reclamation, *Trans. of the World Symposium on Solar Energy*, 1956, Oct

Oswald, W.J., H.B. Gotaas, C.G. Golueke, and W.R. Kellen, Algae in Waste Treatment, *Sewage and Industrial Wastes*, 1957, Apr, v.29, n.4

Oswald, W.J., H.B. Gotaas, and H.F. Ludwig, Photosynthetic Reclamation of Organic Wastes, *The Scientific American*, 1954, Dec, v.79, n.6

Oswald, W.J. and N. Grisanti, Protein From Algae, Presented at Session on Processes for New Protein Foods, *AIChE Symposium Series*, AIChE National Meeting, Kansas City, Mo, 1976, Apr, v.74, n.181

Oswald, W.J., E.W. Lee, B. Adam and K.H. Yao, New Wastewater Treatment Method Yields A Harvest of Saleable Algae, *World Health Organization Chronicle*, 1978, v.32, p.348-350

Oswald, W.J. and H.F. Ludwig, The Role of Algae in Sewage Oxidation Ponds, *The Scientific Monthly*, 1952, Jan, v.74, n.3

Oswald, W.J., G. Shelef, and M. Sabanas, An Improved Algatron Reactor for Photosynthetic Life Support Systems, *1968 Proceedings of the Institute of Environmental Science*, 14th Annual Tech. Meeting, April-May

Oswald, W.J., G. Shelef, and P.H. McGauhey, Algal Reactor for Life Support Systems, *Journal of the Sanitary Engineering Division*, 1970

Oswald, W.J., R.A. Tsugita, R.C. Cooper, and C.G. Golueke, Treatment of Sugarbeet Flume Waste Water by Lagooning: A Pilot Study, *Journal of the American Society of Sugar Beet Technologists* , 1969, v.15, n.4

Palz, W., P. Chartier, and D.O. Hall (eds), *Energy From Biomass*, Applied Science Publ., Barking, Essex, England, 1980, Nov

Parker, D.S., Performance of Alternative Algae Removal Systems, University of Texas Water Resources Center Symposium, Austin, TX, 1975

Payer, H.D., et al., Environmental Influences on the Accumulation of Lead, Cadmium, Mercury, Antimony, Arsenic, Selenium, Bromine, and Tin in Unicellular Algae Cultivated in Thailand and in Germany, *Chemosphere*, 1976, n.6, p.413-418

Phillips, N., and A. Myers, Growth Rate of Chlorella in Flashing Light, *Plant Physiology*, 1953, v.28, p.152-162

Powell, R.C., E.M. Nevels, and M.E. McDowell, *Journal of Nutrition*, 1961, v.75, p.7

Raper Jr., C.D., Plant Growth in Controlled Environments in Response to Characteristics of Nutrient Solutions, NASA Contractor Report 166431, 1982

Richmond, A., A. Vonshak and Shoshana (Malis) Arad, Environmental Limitations in Outdoor Production of Algal Biomass, *Algae Biomass*, Elsevier/North-Holland Biomedical Press, 1980, p.65-72

Rose, A.H., *Economic Microbiology*, Academic Press, 1979, v.4

Russ, Space Men Grew Their Own Food, *San Francisco Chronicle*, 1973, Nov

Ryther, J.H., Potential Productivity of the Sea, *Science*, 1959, v.130, p.602-608

Sabanas, M., C.G. Golueke, H.K. Gee and W.J. Oswald, A Miniature System for Ecological Research, *Proceedings of the Institute of Environmental Science*, 17th Annual Meeting, Los Angeles, CA, 1971

San Pietro, A. (ed), *Biochemical And Photosynthetic Aspects Of Energy Production*, Academic Press, New York, 1980

Schlegel, H., G.G. Gottschalk and R. von Bartha, *Nature*, 1961, v.191, p.463

Shelef, G., R. Moraine, A. Meydan and E. Sandbank, Combined Algae Production: Wastewater Treatment and Reclamation Systems, in Schegel and Barnes (eds), *Microbial. Energy Conversion*, Goltze, Goettingen, 1976, p.399-414

Shelef, G. and C. Soeder (eds), *Algal Biomass*, Elsevier/North Holland Biomedical Press, 1980

Shelef, G., E. Sandbank, Dubinsky, B. Hepher, and A.M. Wachs, Utilization of Solar Energy in a Combined Wastewater Treatment and Algae Protein Production System, *Proceedings of the 5th Conference*, Israel Ecological Society, 1974

Shepelev, Ye. Ya., Biological Systems for Human Life Support (Review of Research in the USSR), 1979 (Translated from Russian) NASA Technical Memorandum TM-76018,

Smith, Dori, Spirulina: Facts Behind the Fad, *Whole Life Times*, 1981, Sep/Oct, n.14, p.4-38

Smith, F. and R. Montgomery, *The Chemistry Of Plant Gums And Mucilages*, Reinhold, New York, 1959

Soeder, C.J., The Scope of Microalgae for Food and Feed, in Shelef and Soeder (eds), *Algae Biomass*, Elsevier/North-Holland Biomedical Press, 1980, p.9-20

Soong, Pinnan, Production of Development of Chlorella and Spirulina in Taiwan, in Shelef and Soeder (eds), *Algae Biomass*, Elsevier/North-Holland Biomedical Press, 1980, p.97-121

Stewart, W.D.P. (ed), *Algal Physiology And Biochemistry*, University of California Press, Berkeley, 1974

Stone, R.W., D.S. Parker and J.A. Cotteral, Upgrading Lagoon Effluent for Best Practicable Treatment, *Journal Water Pollution Control Federation*, 1975, v.47, n.8, p.2019-2042

Switzer, Larry, Spirulina: A Unique Food Choice, *Runner's World*, Mountain View, California, 1982, Aug, v.17, n.8, p.41-88

Tamiya, H., Mass Culture of Algae, *Annual Review of Plant Physiology*, 1957, v.8, p.309-334

Tamura, E., et al, Nutritional Studies on Chlorella: Human Excretion and Absorption of Decolorized Scenedesmus, *Annual Report National Institute Nutrition*, Japan, 1958, v.11, p.31-33

Terskov, I.A., B.G. Gitel'zon, G.V. Kovrov, et al., Closed System: Man-Higher Plants (Four Month Experiment), 1979 (Translated from Russian) NASA Technical Memorandum TM-76452,

Tibbitts, T.W. and D.K. Alford, Use of Higher Plants in Regenerative Life Support Systems, Ames Research Center, NASA, Grant NSG-2405, 1980

Toerien, D.F. and J.U. Grobbelaar, Algal Mass Cultivation Experiments in South Africa, in Shelef and Soeder (eds), *Algae Biomass*, Elsevier/North-Holland Biomedical Press, 1980, p.73-80

Trubachev, I.N., V.A. Barashkov, G.S. Kalacheva and Yu. I. Bayanova, Single-Celled Algae As Potential Source of Food Raw Material in Kordyum (ed) *Role Of Lower Organisms In Recycling Of Substances In Closed Ecological Systems*, 1979, p.206-209 (Translated from Russian) NASA Technical Memorandum TM-76484

Van Vuuren, L.R.J. and F.A. Nan Duren, Removal of Algae From Wastewater Maturation Pond Effluent, *Journal of Water Pollution Control Federation*, 1965, v.37

Venkataraman, L.V., W.E. Becker and T.R. Shamala, Studies on the Cultivation and Utilization of the Alga, *Scenedesmus Acutus As a Single-Cell Protein*, *Life Sciences*, 1977, v.20, n.2

Venkataraman, L.V., B.P. Nigam, and P.K. Ramanathan, Rural Oriented Fresh Water Cultivation and Production of Algae in India, in Shelef and Soeder (eds), *Algae Biomass*, Elsevier/North Holland Biomedical Press, 1980, p.81095

Vincent, W.A., Algae for Food and Feed, *Process Biochemistry*, 1969, v.4, n.45

Waldroup, Park W., Microorganisms As Feed and Food Protein, in Altschul and Wilcke (eds), *New Protein Foods*, Academic Press, 1981, v.4, Part B, p.228-249

Washen, J., Unusual Sources of Protein for Man, Critical Review, *Food Science and Nutrition*, 1975, v.6, n.1, p.77

Weisman, J.C. and J.R. Benemann, *Applied Environmental Microbiology*, 1977, v.33, p.123-131

Wolverton, B.C., Higher Plants for Recycling Human Waste Into Food: Potable Water and Revitalized Air in a Closed Life Support System, ERL Report, NASA, 1980, n.192